CHASING
THE SHADOW

An Observer's Guide to Eclipses

CHASING THE SHADOW

by Joel K. Harris and Richard L. Talcott

from the publishers of ASTRONOMY magazine

KALMBACH BOOKS

Acknowledgments

We would like to thank all the people and organizations who helped us put this book together: F. Richard Stephenson, Joseph E. Kennedy and the University of Saskatchewan Archives, Susan Kitt and the Hayden Planetarium Archives, Ed C. Krupp and Griffith Observatory, Gibson Reeves, Jay Anderson, Carter Roberts, Jay Pasachoff, Stephen J. Edberg, Ernie W. Piini, Alan W. Gorski, and Lief J. Robinson; Tom Hunt, Steve Davis, and Phil Kirchmeier in the Kalmbach Art Department; Mary Algozin, Sabine Beaupré, and Terry Spohn in the Books Department; and last but certainly not least, all the photographers who so graciously allowed us to use their photos.

Dedication

To my late father, I. M. "Bucky" Harris, who fostered my interest in astronomy 35 years ago, and Donald Menzel of Harvard Observatory, who showed me the way to participate in an eclipse on my first eclipse expedition, to Mauritania in June 1973.

Joel Harris

To my parents, for their love and support through the years, and to Evelyn, who makes everything I do seem worthwhile.

Richard Talcott

Library of Congress Cataloging-in-Publication Data

Harris, Joel K.
    Chasing the shadow : an observer's guide to solar eclipses / by
Joel K. Harris and Richard L. Talcott.
    p. cm.
    Includes bibliographical references and index.
    ISBN 0-913135-21-6

    1. Solar eclipses. I. Talcott, Richard L. II. Title. III.
Title: Observer's guide to solar eclipses.

QB541.H37 1994   523.7'8
                               QBI94-679

CONTENTS

INTRODUCTION

A familiar yellow globe hangs in our sky each day, bringing us light and warmth. We curse the Sun in the summer when it's too hot and then curse it again in winter for deserting us. Despite our complaints, we rely on the Sun's familiarity and constancy, knowing that it will always be there, always looking the same. Maybe that's part of the reason eclipses fascinate us so. For a few brief moments, our familiar Sun is transformed into another object entirely. It's not that staid, yellow sphere, but a dynamic object with fiery red tongues of flame leaping from its surface and an eerie, opalescent outer atmosphere that stretches far beyond the limits of what we've seen before.

To see what the Sun is really like, you can't look at it as it appears day in and day out. You must see it when its staid outer veneer has been stripped away by the Moon. Of course, this Sun doesn't show itself to just anyone— you typically have to travel great distances to see it, and even then you get only a brief, tantalizing view.

WHY GO?

Getting that brief view is often a time-consuming and expensive proposition. Veteran eclipse chasers think nothing of spending their hard-earned money and vacation time to visit a remote region of the globe for a few minutes of totality. But at least many people nowadays have the means to do so. Until the last couple of centuries, eclipse observers had to wait for the eclipse to come to them—a daunting task, considering that any spot on Earth witnesses a total eclipse only once every 300 years, on average. This no doubt fueled some of the fear ancient people felt at seeing a total eclipse—most never saw more than one in their life. To take the familiar life-giving Sun away, even if for only a few minutes, was bound to frighten people who were superstitious by nature.

WHAT'LL I SEE?

There are three main types of eclipse: During a partial eclipse the Moon passes in front of the Sun but doesn't block it completely from view. An annular eclipse occurs when the Moon passes directly in front of the Sun's disk but is too small to block all the Sun's light, so that a ring of sunlight remains visible. In a total eclipse the Moon blocks the Sun entirely from view. The beauty of a partial, annular, or total eclipse is what drives people to chase the Moon's shadow. We discuss the three main types of eclipse and the conditions under which they arise in Chapter 2.

Once you've made the decision to go chasing after eclipses, you need a detailed plan for doing so. In this book, we guide you through the often laborious process of figuring out what eclipses make the most sense to see and how to go about seeing them. You need to consider not only the length of the eclipse but also weather prospects along the eclipse path, how accessible various locations are, and the political climate of the areas along the eclipse track. Then you have to decide whether to plan the trip yourself or leave that to an experienced tour guide.

Of course the chief excitement of an eclipse is actually viewing it! We tell you what phenomena to look for during an eclipse and how to view the eclipse safely. At every eclipse, it seems, dozens of people damage their eyes by not taking proper precautions. You'll read all about prominences, the corona, shadow bands, and the 360° twilight. And because many people love to bring home photographic souvenirs of their eclipse trips, we give you a detailed guide on how to photograph an eclipse.

WHEN CAN I SEE IT?

We wrap up the book with an in-depth look at every total and annular eclipse for the remainder of the 1990s. Beginning with a spectacular annular eclipse that cuts a wide swath across the heart of North America on May 10, 1994, and ending with a total eclipse that passes through a significant chunk of Europe on August 11, 1999, the decade offers eclipse chasers four annular and five total eclipses. We give charts for each eclipse and recommend areas that offer good weather prospects for seeing the eclipsed Sun.

So come with us as we explore one of nature's most awe-inspiring sights. And get ready for a decade's worth of memories from the next batch of spectacular eclipses.

Partial phase of eclipse, July 11, 1991, Waikoloa, Hawaii. *Bill and Sally Fletcher.*

8

1

THE ECLIPSE ARCHETYPE

And I will show wonders in the heavens and in the earth, blood,
and fire, and pillars of smoke. The sun shall be turned into
darkness, and the moon into blood. . . .

—Joel 2: 30–31

Eclipses have commanded humankind's attention, imagination, and yes, fear, for as long as people have gazed into the skies above them. Every culture has some type of legend, word, or explanation for the periodic disappearance of the Sun, our stellar companion. This becomes particularly noticeable as you travel across the globe in pursuit of the Moon's shadow. With few exceptions, those living in the path of a total eclipse don't ask for explanations— local knowledge has already provided a reason for what is about to occur. One of the more striking examples of eclipse awareness we encountered came during the first expedition one of us [Harris] participated in, pursuing the eclipse of June 30, 1973, in northwestern Africa.

The expedition, operated by Educational Expeditions International (EEI, now known as Earthwatch, Inc.), was based in central Mauritania, a country on the western edge of the Sahara Desert. For three weeks, nearly 100 hardy individuals camped out, braving daytime temperatures in excess of 120° Fahrenheit as they awaited the seven minutes of umbral passage soon to envelop them.

Shortly after we arrived, a nomadic Mauritanian and his camel entered our

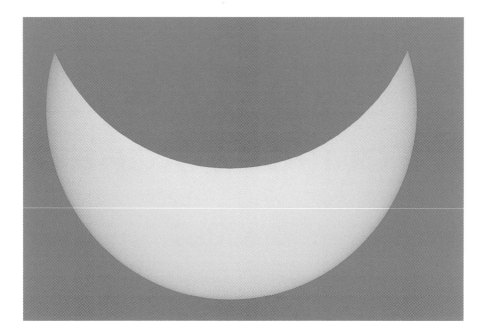

"When they cross, there is an eclipse." Sights such as this one, photographed in March 1988 from Borneo about halfway to totality, often brought on fear and political upheaval in the ancient world. *Jacques Guertin.*

encampment. The Bedouin was seeking treatment for a large sore on his right leg. A member of EEI's support staff, Dr. Goodman, applied some antibiotic cream to the wound and then dressed it. Afterward, several of us sat around consuming hot tea prepared the Arabic way—not strong, but extremely sweet. Jane Fisher, another EEI staffer, asked the nomad in French whether he knew that a solar eclipse was to occur soon. He said yes, many of his fellow tribesmen had heard about it from others and from radio reports.

Fisher then asked if he knew what caused the eclipse, and he nodded. He took a thin twig and smoothed some of the Sahara's sand with his parched hand. He then drew two lines that crossed in the middle. "This is the path the Sun takes," he said as he retraced one line. "This is the path the Moon takes," he said as he traced the other. "When they cross, there is an eclipse."

Fisher asked our visitor if he was at all worried about the impending event. The tribesman said, "No, we are not afraid of the eclipse because the

American scientists are here making it happen correctly, and so there is no reason for fear."

Dr. Edwin C. Krupp, archeoastronomer and director of the world-famous Griffith Observatory in Los Angeles, California, has said, "For all practical purposes, the Sun and Moon are agents of world order, symbols of the world and the revelation of that order. They are sources of power. Then you get these occasions that are just bonkers. . . . Historically, a perfect meeting between the Sun and Moon means an eclipse. And eclipses mean trouble."[1]

ECLIPSES IN ANTIQUITY

One of the earliest written records of an eclipse of the Sun, on May 3, 1375 B.C., was found in the city of Ugarit in old Mesopotamia. It read:

> On the day of the new moon, in the month of Hiyar, the Sun
> was put to shame, and went down in the daytime, with Mars
> in attendance.[2]

From at least the time of this ancient record onward, peoples around the world have considered eclipses to be portents of ill. The well-known story about the royal Chinese astronomers Hsi and Ho, who failed to predict an eclipse and paid for this transgression with their lives, is a classic example of the importance emperors and other heads of state placed on accurate prediction of eclipses.

Another example of how rulers during the last two millennia closely watched for eclipses and were wary of how the events could affect their rule is given in a paper by P. Miller of the Jet Propulsion Laboratory and F. R. Stephenson of the University of Newcastle-upon-Tyne, England:

> . . . The astronomer Royal Hsü Chih submitted to the Throne
> a warning of a solar eclipse and requested that he and the
> Prime Minister should be granted leave to sacrifice at the
> Imperial Observatory in order to avert the impending
> calamity. The Emperor replied, "I have heard that when the
> Emperor's actions are faulty, the heavens are alarmed and
> manifest portents as warnings to enable him to repent. Thus

when the Sun and Moon are veiled or eclipsed one can infer irregularities in the administration."[3]

Early eclipse observers were aware of the physical dangers in viewing partial eclipses directly. The Islamic astronomer al-Biruni notes that:

> . . . the faculty of sight cannot resist the Sun's rays, which can inflict a painful injury. If one continues to look at it, one's sight becomes dazzled and dimmed, so it is preferable to look at its image in water and avoid a direct look at it, because the intensity of its rays is thereby reduced and one can look at its disk.[4]

Thus, even the most ancient eclipse observers were well aware of the potential hazards of eclipse viewing. They knew that one must never look

The root word for "eclipse" is generally thought to be the Greek word, "ekleipsis," meaning "abandonment." Here, in the moment of totality, it's easy to see how early witnesses may have felt terror. *Ron Lee and Roberta Ewart.*

directly at the Sun while observing (at least prior to the advent of Mylar and other safe forms of solar filter materials).

Upon investigation, the depth and breadth of the symbolism of eclipses in the history of civilization become apparent. The word eclipse itself hints at something negative. The most commonly accepted genesis of the word is that it stems from the ancient Greek word ekleipsis, meaning "abandonment" or "omission." Certainly, a word that carries this overtone must denote something unsavory. As the historian Ibn al-Athir stated in April 1176:

> In this year the Sun was eclipsed totally and the Earth was in darkness so that it was like a dark night and the stars appeared. That was the forenoon of Friday the 29th of Ramadan at Jazirat Ibn 'Umar, when I was young and in the company of my Arithmetic teacher. When I saw it I was very much afraid; I held on to him and my heart was strengthened. My teacher was learned about the stars and told me, 'Now you will see that all of this will go away,' and it went quickly.[5]

ECLIPSE MYTHS AND LEGENDS

Stories and tales that seek to explain eclipses arise in virtually every culture on Earth. And almost every one of them centers around some type of celestial creature that has chosen the time of the eclipse to have a stellar meal. The ancient Chinese believed a huge dragon was consuming the Sun, a concept also held by the Greeks and Romans and later on by the Balt people of northern Europe. In fact, the Mandarin Chinese word for eclipse, "shih," means "to eat."[6]

Images of dragons eating or chasing the Sun (or Moon) about the sky are well ingrained in the history of astronomy, but similar images endure to the present day. The symbol for the ascending and descending nodes of the Moon's orbit is the Greek letter omega (Ω). Coincidentally, the nodes are those places in the Moon's orbit where the geometry permits a lunar or solar eclipse (see Chapter 2). It takes only a little imagination to see a resemblance between this letter and a dragon or serpent, again reinforcing the concept that during an eclipse, someone or something is eating the Sun or Moon. The

The "bite" taken by the Moon led to the rise of many tales of celestial animals who pursued and ate the Sun. Always, however, the Sun emerges triumphant. *Ronald Schmidli.*

theme of some form of supernatural animal, being, or force consuming the Sun recurs in many cultures. Vampires (in Siberia), werewolves (Serbia), giant frogs (Vietnam), jaguars (Paraguay and Argentina), and oversized dogs (Bolivia) are all alleged culprits that cause the periodic absence of the Sun or Moon from the sky.[7]

One of the most engaging tales about how eclipses occur is a myth prevalent in Indonesia. Natives of the island Bali relate the following version of a story derived from the Indian Hindu epic poem the Mahabharata.

The evil antagonist, Kala Rau (Rahu in the Indian version), is jealous of the gods that inhabit Nirvana, who are all-knowing and immortal. Kala devises a plan to live as a god by disguising himself as a woman and serving the magic elixir soma to the gods at a heavenly banquet.

During some commotion at the banquet, Kala Rau uses the diversion as an opportunity to sneak a swig of the soma and thereby grab his chance at immortality. However, someone tips off Vishnu (one of the Hindu deities), and the jig is up. Quicker than you can say alakazam, Kala is decapitated. The soma barely gets past Kala Rau's throat when his headless body dies. His head is the only part of his being that the potion has time to immortalize. Kala becomes so angry that from then on he chases the deities

14

of the Sun and Moon across the heavens, hoping to catch and eat them.

From time to time, Kala Rau succeeds in his quest and is able to consume some or all of the Sun or Moon. However, the hole in his neck where he was decapitated permits him to ingest either body only for a short time before it again reappears in its full glory. Some say the extreme heat of the Sun keeps him from swallowing it.

ECLIPSES DEPICTED IN WRITING

Since the beginning of modern times, particularly in the 19th and 20th centuries, people have sought to describe the splendor of totality in writing. This effort proved especially valuable in the days before photographs and motion pictures were generally available to the public. Additionally, prior to the latter part of the last century, only a few scientists and their assistants (or fortunate family members) ever journeyed into the path of a total eclipse. To do so, they had to travel for many weeks and endure extremely difficult field conditions. And even those who did venture to distant regions lacked the accurate weather forecasts that would improve their chances of seeing an eclipse. The bottom line was that few people had the chance to see a total eclipse, let alone describe what it looked like to non-observers.

The most common outlet for writing about the eclipse experience was the summary report that followed the completion of every eclipse expedition. Teams of astronomers and staff, dispatched by the U.S. Naval Observatory, the Royal Astronomical Society, trading companies, private foundations, universities, and other philanthropic or educational organizations, were some of the first to start the now long-standing tradition of telling about their successes (and often failures) in chasing the lunar shadow across the globe.

William Coldwell and William Buckingham wrote for the *Nor'Wester,* a newspaper published in the town of Fort Garry, Manitoba, now known as Winnipeg. They accompanied the famous solar astronomer Simon Newcomb on an expedition funded by the Hudson's Bay Company to observe the total eclipse of July 18, 1860. They remained, however, outside the path of totality in Upper Fort Garry to witness a deep partial eclipse.

> At the commencement of the obscuration, the sky was overcast, with heavy masses of cloud in the east, and there was

much reason to fear that the celestial phenomenon would not be at all apparent hereabouts. But a brisk gale of wind having scattered the clouds, shortly before six o'clock the sun became visible to the eager gaze of thousands, and again astronomic prediction was verified. The black shadow had eaten its way a considerable distance into the surface of the bright orb, and slowly but steadily the darkness appeared to extend itself over that dazzling surface. What a scrutiny the great change was attracting from all quarters of the earth! What an array of telescopes were eagerly searching the blue vault above during those precious moments![8]

As luck would have it, Newcomb ought to have remained with Coldwell and Buckingham at Upper Fort Garry. The following passage, written by S. H. Scudder for the publication *The Winnipeg Country* in 1886, recounts the ultimate failure of Newcomb's party:

The evening was rainy, and the astronomers retired without any bright hopes of successful observations the following day. The early morning threatened that their fears would be well-founded; but preparations were nevertheless made to take whatever observations might be possible. . . . But these arrangements were all to no purpose so far as any observations of the phenomena of the total eclipse was concerned. The sun remained covered with thick clouds during the minute of totality and until some time after the flash of returning light had struck the eye.[9]

The disappointed Scudder made no secret of his feelings about being deprived of seeing totality:

Three thousand miles of travel occupying five weeks to reach by heroic endeavor the outer edge of the belt of totality, to sit in a marsh and view the eclipse through the clouds. . . .[10]

". . .everything terrestrial took on a cold iron hue. . . ." Within a few minutes an eclipse transforms midday to evening; the air cools suddenly, birds take flight for home, a breeze rises. *Barney Magrath and Oluf Nielsen.*

This was no piece of journalistic hyperbole. The Hudson's Bay Company expedition had endured very difficult conditions just to reach the path of totality. A sample entry from Newcomb's personal diary indicates the traveling conditions the eclipse party had to endure:

> Started at 6+ AM. Pelican Lake dotted with islands. Must get out and walk at the end of every mile. Gager accompanied us. At dinner no horses to change, so hitched to two road wagons. Got badly stuck twice, with water over boots. Last time at dusk with threatening thunderstorm. Arrived at Brekenridge at 11½ PM, spread a tarpaulin on the ground and went to sleep.[11]

Not every eclipse expedition had such disappointing results, however. The total eclipse of May 28, 1900, which moved from the southeastern coast of the United States, across the Atlantic Ocean, through Spain, across the Mediterranean, and into Algeria, was seen by many expeditions stationed along the path.

One of the groups observing the event under the auspices of the British Astronomical Association was stationed near Navalmoral, Spain. A beautiful and moving depiction of seeing totality appears here, as related by T. Weir, a member of the party who was a fellow of the Royal Astronomical Society:

> . . . the semi-darkness, for there was no real blackness, came on suddenly, and during totality, computed to last 1m 28s., everything terrestrial took on a cold iron hue, altogether different from the gloom of evening. The distant town and more distant mountains were almost blotted out from view, whilst in the heavens above round the moon's black disk, as if by the touch of a magician's wand, there flashed out the corona in grandeur of form and of pearly whiteness. Mercury, too, in close proximity, shone with the brilliance of a miniature sun, and enveloping the whole was a halo of soft white light; a spectacle whose unique beauty words fail utterly to describe.[12]

Were it not for the fortunate coincidence that we live on the one and only planet in our solar system where, thanks to some lucky geometry, the Sun and Moon just happen to be almost the same apparent size in the sky, total solar eclipses would be rendered impossible. In fact, we are even more fortunate than we realize. The constant tidal motion and gravitational effect our oceans and seas have upon the orbit of the Moon are inexorably pushing Luna ever farther away from our world. At some time in the dim future, our satellite will be too distant from us to create a total solar eclipse, and we will be left with only partial and annular type events to observe.

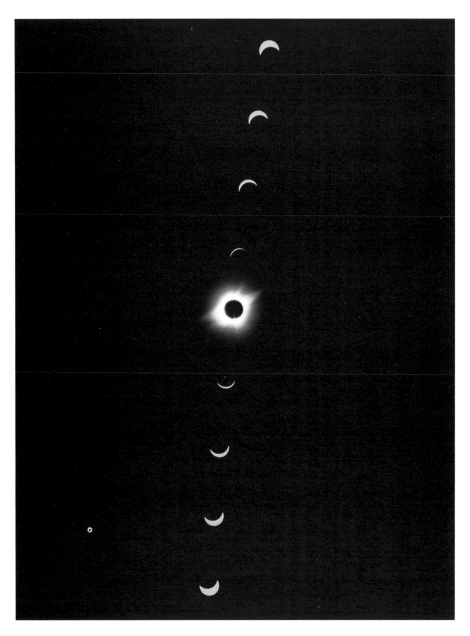

This multiple image sequence depicts all the stages of the most spectacular of solar eclipses, a total eclipse. One image was shot every eight minutes. *William P. Sterne, Jr.*

2

A SOLAR ECLIPSE PRIMER

A solar eclipse is deceptively simple. The Moon moves directly between the Sun and Earth. However, like many celestial events, this simple geometric event is more complex than it looks. Solar eclipses come in three varieties— total, partial, and annular. Let's look first at the different types and the circumstances that cause each of them.

MONTHS AND NODES

Each month the Moon orbits once about our planet. Indeed, the month as a unit of time and the word "month" itself derive directly from this fundamental motion. But if the Moon passes between the Sun and Earth each month, why doesn't a solar eclipse occur every month? The reason is that the orbit of the Moon is inclined ever so slightly to the orbit of Earth. This five-degree tilt means that the Moon crosses Earth's orbital plane only twice a month. To complicate matters further, the points at which these crossings occur move constantly from east to west—"regressing," as astronomers say.

This westward motion is caused by the varying lengths of time known as a month. The first type of month—the synodic month—is the period between successive New or Full Moons. It is 29.53 days long. Second is the sidereal month—the time for the Moon to return to exactly the same position in the sky relative to the stars. This lasts about two days less than a synodic month, or 27.32 days. Third is the eclipse, or draconic, month—the period between two successive orbital crossings. This month lasts 27.21 days.

The two points at which the Moon's orbit crosses Earth's path around the Sun (and thus the Sun's apparent path in our sky) are called "nodes." The one where the Moon crosses from north to south is called the descending node, and the one where the Moon crosses from south to north, the ascending node. Generally, the Moon crosses one of the nodes at or near a point where the Sun appears in the sky about twice a year. It can happen more often, but only rarely. For a solar eclipse to occur, the Moon must be at the New phase when it crosses one of the nodes. Remember, a solar eclipse happens only when the Sun, Moon, and Earth are aligned in that order. If the Moon crosses one of the nodal points when it is opposite the Sun's location in the sky—that is, when the three bodies are aligned in the order Sun, Earth, and Moon—the geometry produces a lunar eclipse. At these times Earth's shadow is cast upon the Moon, and the Moon can appear any color from dark grey to a bright orange-red.

The final factor governing the type of solar eclipse is the distance of the Sun and Moon from Earth. The orbits of the Moon around Earth and Earth around the Sun are not precise circles, but ellipses. So the distances of both the Sun and the Moon—and thus their apparent size in our sky—vary throughout the course of the year and month, respectively.

Contrary to popular belief, Earth lies nearest the Sun in January. Many people assume the Sun is closest during the hot summer months, but Earthly temperatures are governed more by the angle at which sunlight strikes Earth than by the Sun's proximity. In January, Earth lies 3 percent closer to the Sun (about 3 million miles closer) than in July. The Moon's distance from Earth varies even more dramatically. It ranges from 217,000 miles to 249,000 miles—a 15 percent difference. The time between two closest approaches of the Moon to Earth is yet another kind of month—an anomalistic month, equal to 27.55 days

THREE TYPES OF SOLAR ECLIPSES

Whether an eclipse is total, partial, or annular depends on two things: how close the New Moon comes to a nodal point, and the relative size of the Sun and Moon in the sky. The most common type of eclipse occurs when the

Normally made invisible by the much brighter light of the photosphere, the Sun's corona glows eerily during a total eclipse. This stage of an eclipse lasts only a few seconds to a few minutes. *David M. Beraru.*

Moon lies slightly off a nodal point, so that the lunar disk only partially covers the Sun. About 35 percent of all solar eclipses are such partial eclipses. They can be seen from wide areas, so a great many people have seen at least one. In a partial eclipse, the Moon appears to take a bite out of the Sun. The amount of the bite can be anywhere from a barely perceptible 1 percent to a near-total 99 percent.

The next most common solar eclipse is an annular, or ring, eclipse. The name comes from the Latin word "annulus," meaning ring. During an annular eclipse, a thin ring of sunlight remains visible around the Moon's disk. Though the Moon lies directly in front of the Sun, it is too small to cover the Sun's disk. The Moon is too far away from Earth and thus appears small, or the Sun is too close to Earth and thus appears large—or both may be the case at once. A ring of the Sun's photosphere remains visible, washing out the more subtle light of the Sun's corona and chromosphere and spoiling the spectacle of totality. About 32 percent of all solar eclipses are annular.

The most spectacular type of eclipse is a total eclipse, which happens about 28 percent of the time. During such events the sky grows dark, stars appear, and the horizon is bathed in an eerie, 360° twilight glow, while the Sun's pearly white corona and ruddy chromosphere come into view.

An even rarer event, called an annular-total eclipse, occurs about 5 percent of the time. In this unusual event, annularity briefly and barely becomes totality, with the total phase lasting for only a second or so.

BRIEF MOMENTS OF GRANDEUR

You rarely have time to savor an eclipse. The total phase of a total eclipse can last no longer than seven and a half minutes, and most are considerably shorter than that. Annular eclipses typically last a few minutes, though annularity can last up to twelve minutes under the right circumstances. Partial eclipses have the greatest range: they can be finished in a few seconds or last for over two hours. As an added bonus, all annular and total eclipses are preceded and followed by lengthy partial phases.

Several factors control the duration of an eclipse. We've already touched on one—the relative sizes of the Sun and Moon. All other things being equal, a total eclipse lasts longer the larger the Moon appears relative to the Sun. An annular eclipse lasts longer the smaller the Moon appears relative to the Sun.

Another element dictating the length of an eclipse is the latitude on Earth where the eclipse takes place. Every point on our world rotates on its axis once every 24 hours. But the actual speed you travel varies depending on your geographic latitude. If you live on the equator for example, Earth moves at a rate of about 1,000 miles per hour toward the east. If you live at say, 40° north or south latitude, the world moves around 600 miles per hour to the east. At the poles there is no eastward motion at all.

The lunar shadow, or umbra, travels from west to east across the Earth's surface. The faster Earth rotates at your location, the more this slows the speed of the umbra along the eclipse track. In other words, Earth's eastward spin reduces the velocity of the Moon's shadow and allows you more time within it. It's much like racing along the highway, trying to catch a car going

THE SOLAR ECLIPSE PRIMER

Partial eclipses are not only the most common, they can be seen from the widest area, so this is the kind of eclipse people are most likely to see. Note the sunspots visible on the lower half of the Sun in this photo. *Mark Coco.*

faster than you. If you accelerate enough to match the other car's speed, you'll be able to keep up as long as the race continues.

Sadly, most of us can't arrange to travel at nearly 1,700 miles an hour to reduce the lunar umbra's speed to zero. However, this feat was nearly achieved in 1973, when a prototype of the Concorde supersonic jet raced along with an eclipse shadow for 70 minutes!

The final two factors that affect the length of the annular or total phase are your position along the west-to-east eclipse path and your proximity to the center of the path of annularity or totality. If you're located near the point where the Moon's shadow first touches Earth at sunrise or finally leaves Earth at sunset, the event will take place much faster. Near sunrise or sunset the umbra (or antumbra, in the case of an annular eclipse) is nearly tangential to Earth's surface, and its speed is correspondingly faster due to the oblique angle at which it strikes Earth. As the Sun rises higher in the sky, the speed of

An annular eclipse occurs when the diameter of the Moon is smaller than that of the Sun, which means the Moon is farther from Earth—or the Sun is closer—than during a total eclipse. *Phil Lohmann.*

the lunar shadow slows down. Near the midpoint of the eclipse track, Earth is slightly closer to the Moon's surface. The Moon thus appears slightly larger than at either end of the path, contributing slightly to a longer totality.

During an annular eclipse, however, the closer you lie to the midpoint of the track, the closer you move to the vertex of the antumbral cone. The diameter of the shadow thus becomes smaller, and annularity is therefore slightly shorter.

In both types of eclipse, the closer you get to the center of the north or south limit of an eclipse path, the longer the total or annular phase.

ECLIPSES IN CYCLES

Solar eclipses show a symmetry dictated by the laws of mathematics and celestial mechanics. It turns out that 223 synodic months is very nearly equal to 242 draconic months and to 239 anomalistic months. Thus, approximately every 18 years, 11.3 days, the Sun, Moon, and Earth return to the same rela-

At the height of a total eclipse Earth's horizon is bathed in twilight in every direction. This eerie condition adds to the unforgettable beauty of the event. *Carter Roberts.*

tive position. This coincidence defines a pattern known as the saros; it represents a cycle or repetition of one particular eclipse track. An outstanding example of a saros is saros 136. This family of events brought the world the unusually long total eclipses of 1919, 1937, 1955, 1973, and most recently, the big one of 1991.

Some fast arithmetic shows that the listed dates are 18 years apart. And if you looked at the actual calendar dates of the eclipses, you'd see that each occurred 11.3 days later than the previous one in the series. The additional one-third of a day in a saros causes the path of the eclipse to shift 120° to the west of its predecessor's position. That's because Earth has rotated an extra third of a day beyond where it was for the previous member of the saros.

A saros is born every 29 years—a period called an "inex." When a saros starts, it begins at one of Earth's poles as a partial solar eclipse. As the cycle progresses, the eclipse paths move toward the equator and the Sun plunges deeper into the Moon's embrace, creating greater partial eclipses. The eclipses eventually become annular and then total before they start growing shorter again. As the saros winds down, the eclipses again become only partial, and their paths migrate toward the pole opposite the one where they originated. At any one time, approximately 42 saroses are in action. Some are beginning, some ending, and some just reaching their greatest duration.

A Sight to Behold

Viewing an eclipse brings on a range of emotions. A partial eclipse is like eating a bread-and-butter sandwich. You know you've had something to eat, but you really aren't very satisfied. Usually the Sun has only a modest piece taken out of it, and you don't notice much change in your surroundings.

An annular eclipse is more dramatic, particularly if it occurs at sunrise or sunset, or if the Sun and Moon are nearly the same apparent diameter. The sky gets noticeably darker, as if it were a very cloudy day; the temperature may drop noticeably if the day is warm. Nevertheless, a significant amount of sunlight remains, and your surroundings appear pretty normal.

A total eclipse is stupefying. Words cannot adequately describe it. No photo can capture the full dynamic range of light and color it displays. To many, it is the single most dramatic natural phenomenon you can observe on Earth. Even if clouds block the Sun and prevent you from seeing the Sun's

corona or any promi-
nences, the show is awe-
some. The sky grows
much darker than during
daytime, similar to mid-
twilight. The air feels
markedly cooler. Bright
stars and planets appear
in the sky. Animals and
humans react strangely.
Often a modest wind, or
"eclipse breeze," comes
just before totality. And
the trademark 360° sun-
set is visible along the
entire horizon. Seeing a

**Often there are too many changes on the ground
and in the sky for one witness to take in during the
few minutes of totality.** *David M. Beraru.*

total solar eclipse may be the closest thing to visiting another planet that anyone living in the 20th century will ever experience. It transforms even familiar surroundings into an alien scene.

Even for one of us [Harris] who has seen nine total eclipses, the experience makes the blood run faster, winds up the emotions like a child's toy, and
sends a chill calm through his entire being. If you keep no other promise to
yourself, you must witness the grandeur of totality. You'll emerge from the
experience a different person.

The Sun's wispy corona extends millions of miles out into space, but it's visible only during a total eclipse. *Huguette Guertin.*

3

WHY THE PURSUIT?

Until the 1930s, a total eclipse of the Sun was the only way to closely and safely observe the normally invisible regions of our closest star—the chromosphere, prominences, and corona. Yet surprisingly, the scientific study of eclipses only got underway about 100 years before then.

On May 15, 1836, the Moon's shadow touched down in southern Scotland. Although it was only an annular eclipse, English astronomer Francis Baily was there to view it. As the Moon nearly covered the Sun, he saw how light from the edge of the Sun was blocked by tall mountains along the lunar limb, or edge, but poured through deep valleys. Baily's description of the beadlike appearance of the Sun's light inspired the name Baily's beads. But more important, Baily's description of the event so inspired others that he helped start the practice of eclipse chasing. Six years later in southern Europe, a total eclipse—particularly the stunning sight of the beautiful and ethereal corona—sparked even greater interest.

Since Baily observed these eclipses, astronomers have followed the Moon's shadow wherever it led. They have mounted expeditions to the far corners of the globe, carrying literally tons of equipment with them, to view the fleeting moments of totality and learn what they could about the Sun.

DECIPHERING THE SUN

Astronomers took a major step in this direction during a European eclipse on July 18, 1860. From observing sites about 250 miles apart, British astronomer Warren De La Rue and Italian astronomer Angelo Secchi used the relatively new technique of photography to record prominences. At the time, many astronomers thought prominences were part of the Moon, but the photos proved otherwise. The prominences appeared the same from both locations, which meant that they could not be as close as the Moon and must belong to the Sun instead.

An even more significant finding came out of the eclipse of August 18, 1868, in India and Malaysia. Armed with the newly invented spectroscope, astronomers scrutinized the Sun's gases in unprecedented detail. The spectroscope breaks light into its component colors and reveals not only the composition of the gases giving off the light but also their temperature and density. The scientists found that most of the Sun is made of hydrogen gas, which we now know is the most abundant element in the universe.

But the spectroscope also revealed an element no one had seen before. During the eclipse and in the months afterward, French astronomer Jules Janssen and English astronomer Norman Lockyer closely examined the spectrum of the solar prominences. Lockyer found a spectral line in the yellow part of the spectrum that corresponded to no known element. He correctly inferred that it came from a new element, which he named helium after the Greek word for Sun, helios.

EINSTEIN'S PROOF

But perhaps the most important observations of a solar eclipse came in the early 1900s. In the first fifteen years of the century, German physicist Albert Einstein developed two startling and wide-ranging theories—the special and general theories of relativity. The special theory, which dealt with the motions of objects traveling at constant speeds, was far easier to understand and could be tested in many ways. But the general theory, completed in 1915, dealt with objects that moved at varying speeds. Einstein described the movement of these objects in terms of how they distorted the geometry of space. In essence, massive objects distort space more than light objects do, so they have a greater effect on the motion of objects passing near them.

At the time, there were few ways to test the validity of general relativity. One method was to see whether light passing near a massive object was deflected. The perfect test would be to see whether light from distant stars

This photo, taken three seconds before third contact (when the Sun first emerges from behind the Moon), reveals a spectacular fiery prominence rising from the Sun's surface. *Jeff Schroeder.*

was deflected by the right amount as it passed near the limb of the Sun. But of course stars in the vicinity of the Sun are invisible because the Sun is so bright—except during a total solar eclipse.

In 1919 British astronomers mounted two expeditions to view the May 29 eclipse and to try to confirm Einstein's prediction. One group, led by Andrew Crommelin and Charles Davidson, headed off to Brazil to photograph the Sun during totality; the other group, headed by Arthur Eddington and Edwin Cottingham, set sail for the island of Principe off the west coast of Africa. Although both groups had to endure threatening weather, they saw totality and got their photographic plates.

When they compared those plates with ones of the same region of sky taken when the Sun wasn't nearby, they found that the stars' light had indeed been deflected, and by just the amount predicted by the theory. The announcement of the confirmation created a huge stir, no doubt partly because British researchers had confirmed a German scientist's predictions just a year after their two countries had fought one another in the Great War.

UNLOCKING MORE SECRETS

Astronomers continued to study the deflection of starlight and aspects of the Sun's outer atmosphere during subsequent total eclipses. But the invention in 1930 of the coronagraph by French astronomer Bernard Lyot ushered in a new era of solar observation. The coronagraph allowed astronomers to study the subtler components of our star without a total eclipse.

After the introduction of the coronagraph, as well as the further refine-

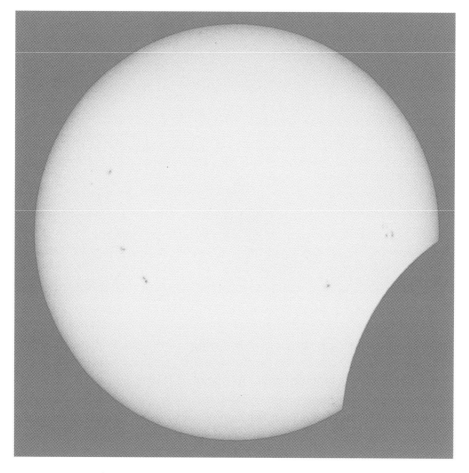

Eclipses offer scientists rare opportunities to measure precisely the Sun's diameter. It is thought that the Sun expands and contracts periodically, as if it were breathing. Sunspots dot the Sun's surface in this photo. *H. J. P. Arnold.*

ment of devices like the Hα, or hydrogen-alpha, filter, it became unnecessary to travel long distances to conduct solar studies. Why, then, have scientists and amateur astronomers continued to chase the Moon's shadow? From a purely technical standpoint, many investigations of the Sun can still be most easily accomplished at the time of a solar eclipse. These include, for example, studies by both professional and amateur astronomers to measure the Sun's diameter to a higher degree of accuracy by making exact timings of contact times during eclipses.

Even though the Sun's diameter is well known to modern astronomers, continuing measurements of its size are still important. This is due to the recent development of the science of helioseismology—the study of the Sun's internal dynamics. It is widely believed that the Sun actually expands and contracts, as if it were breathing, on a periodic cycle. Measuring the diameter of the solar disk at an eclipse provides further data toward confirming or rejecting this theory.

Astronomers like Paul Malley in Texas have specialized their studies even more, with an interest in solar eclipse limb phenomena. Malley intentionally places himself at either the north or south limits of the path of totality in order to extend events like Baily's beads and the appearance of the chromosphere. Other common studies at total solar eclipses include measuring the precise degree of sky darkening throughout totality; observing the shape, extension, and details of the corona; photographing and analyzing shadow bands; and making exact measurements of the drop in temperature that attends totality.

SOMETHING BIGGER THAN A CLOAK

But perhaps the most popular activity during solar eclipses is the simple visual observation of totality itself. As one of us [Harris] has often remarked, the experience of totality is probably the closest thing to visiting an alien planet anyone living in the late twentieth century will ever experience.

The experience of totality can convert even the most skeptical first-time eclipse observer into a rabid, globe-trotting umbral fanatic.

> The whole fleet was in readiness, and Pericles on board his galley, when there happened an Eclipse of the Sun. The

"Where is the difference, then, between this and the other, except that something bigger than my cloak causes the eclipse?" Here, during the moments of totality, the Sun's corona as well as several fiery prominences are visible. *Mark and Tami Croom.*

sudden darkness was looked upon as an unfavorable omen, and threw the sailors into the greatest consternation. Pericles, observing that the pilot was much astonished and perplexed, took his cloak and, having covered his eyes with it, asked him if he found anything terrible in that, or considered it a bad presage? Upon his answering in the negative, he asked: "Where is the difference, then, between this and the other, except that something bigger than my cloak causes the eclipse?"[13]

Plutarch, *Life of Pericles*

Scattered throughout the long history of eclipse watching, even the most dispassionate scientific observers have been profoundly moved upon seeing the approach and commencement of second contact.

But when the Sun, reduced to a very narrow filament, began to throw upon the horizon a feeble light, a sort of uneasiness seized upon all; every person felt a desire to communicate his impressions to those around him. Hence arose a deep murmur, resembling that sent forth by the distant ocean after a tempest. The hum of voices increased in intensity as the solar crescent grew more slender; at length the crescent disappeared and darkness suddenly succeeded light, and an absolute silence marked this phase of the eclipse . . . the phenomenon in its magnificence had triumphed over the petulance of youth, over the levity which certain persons assume as a sign of superiority. . . . A profound stillness also reigned in the air, the birds had ceased to sing. After an interval of solemn expectation, which lasted about two minutes, transports of joy shouts of enthusiastic applause, saluted with the same accord, the same spontaneous feeling, the first reappearance of the rays of the Sun.[14]

Jean Dominique Francois Arago
Eclipse of July 8, 1842

Once people fled eclipses; now they flock to them. Eclipse watching has created a boom business in ecotourism as thousands travel to remote spots to see such spectacular views as this, witnessed in New Guinea in 1984. *Jacques Guertin.*

FROM TERROR TO FASCINATION

We have seen across the ages how people have reacted to the eclipse phenomenon. In ancient times, total, annular, and deeply partial solar eclipses evoked images of terror and foreboding. They seemed to be a threat to the celestial order of things.

People hid indoors from the lunar umbra. Even today, in some remote regions of the world, superstition rules and governments warn the people to remain indoors, or to keep "eyes down on Eclipse Day." Ancient myths warn pregnant women that if they are exposed to an eclipse, their children will be born with dark spots, due to the powerful influence of the Moon's umbra.

Fortunately, most superstitions have now been superseded by factual information about eclipses. As Ed Krupp of Griffith Observatory was moved to remark: "Three centuries ago, people ran away from eclipses. Now they flock towards them. In many ways, eclipse chasing has become one of the leading examples of ecotourism in the late twentieth century, up there with going to Africa to photograph animals in the wild or whale watching."

> To witness a total eclipse of the Sun is a privilege that comes, in general, to but few people. Many live and die without ever beholding one. Once seen, however, it is a phenomenon never to be forgotten. The black body of the Moon standing out, a huge globe, in sinister relief between Sun and Earth, the sudden out flashing glory and radiance of the pearly corona which can be seen at no other time, the scarlet prominences rising from the surface of the hidden Sun to heights of many thousand miles, the unaccustomed presence of the brighter stars and planets in the daytime, the darkness of twilight and the unusual chill in the air, there is something in it all that affects even the strongest nerves and it is almost with a sigh of relief that we hail the return of the friendly Sun.[15]
>
> Isabel Martin Lewis

Today, anyone who wants to plan ahead can be assured of seeing an eclipse under the most favorable circumstances possible, either on land or at sea. *Joe Reed.*

4

PLANNING AND PREPARATION

Unlike a century ago, today's eclipse enthusiasts do not normally have to endure great hardship, physical or otherwise, to place themselves in the path of an eclipse. In the pursuit of the lunar shadow, earlier intrepid astronomers faced constant peril from the weather, human disease, exotic culture shock, and linguistic barriers. The modern eclipse chaser has only to decide upon a particular eclipse to observe, consult a current airline flight schedule, and find the necessary resources to get into the path of the eclipse on the appointed day and time. However, the umbra addict must still take pains preparing for an eclipse trip. An ill-planned trip could leave the eclipse chaser disappointed and disillusioned, vowing never again to hunt for the lunar shadow.

RESEARCHING ECLIPSE CIRCUMSTANCES

Until the middle of the twentieth century, finding out the when, where, and how of any specific solar eclipse presented a formidable conundrum to both amateur and professional astronomers alike. Probably the best-known and most-respected source of general eclipse circumstances was *Canon of Eclipses,* prepared by Theodor von Oppolzer and later translated by Owen Gingerich (Dover Press, 1962). This watershed work covered the central solar eclipses—those that are either total or annular—from 1207 B.C. to A.D. 2161. Close on the heels of von Oppolzer's book was *Canon of Solar Eclipses,* written by Jean Meeus, Carl C. Grosjean, and Willy Vanderleen (Pergamon Press, 1966). This book contains the paths of all the central eclipses of the Sun from 1898 to 2510. Meeus and Hermann Mucke also published a later work, *Canon of Solar Eclipses: -2003 to +2526* (Astronomisches Büro, 1983).

Except for these major works, though, the amateur astronomer or first-time eclipse enthusiast who wanted to ferret out the particulars of a specific eclipse in the future was rather strapped. A basic knowledge of the periodicity of solar eclipses might tell you that the general path of a past solar eclipse would repeat itself every 18 years, $11\frac{1}{3}$ days, in accord with the saros cycle. Unfortunately, knowing the saros number or general shape of a future eclipse's central path wasn't enough for making accurate predictions. Pinpointing the exact circumstances of the eclipse path required detailed calculations of the size, width, and shape of the lunar shadow from the basic geometry of the Sun, Moon, and Earth. One such calculation—the one that determines the radius of the umbra as seen from a particular location—is:

$$r_u = k (D_u - D_t) / D_u$$
Where:
r_u = radius of the umbral shadow
k = mean radius of the Moon (in Earth radii)
D_u = length of the lunar shadow (in Earth radii)
D_t = observer's topocentric distance from the Moon (in Earth radii)

Yet even armed with this degree of detail, it was difficult to get a highly accurate plot of where on Earth the path of a central eclipse would pass.

Commencing in 1949, the United States Naval Observatory in Washington, D.C., provided a tremendous public service to the amateur and professional astronomical communities. Starting in that year, the Observatory regularly published eclipse circulars that were free to the public for the asking. Each described in detail a specific total or annular eclipse. Unfortunately, the fiscal woes of the late 1980s and early 1990s led to austerity measures, and the U.S. government suspended the publication of these highly useful circulars.

Luckily for all of us, one of the chief contributors to the defunct circulars, senior meteorologist Jay Anderson of the Prairie Weather Centre in

The right equipment is vital to a successful eclipse-watching experience, but often the limitations imposed by weight and size restrictions during the journey require some tough choices. *Alice Newton.*

Winnipeg, Canada, has joined Fred Espenak of NASA's Goddard Space Flight Center in Maryland to fill this major gap in astronomical information.

Anderson and Espenak have received government help to continue publishing eclipse circulars under the auspices of NASA, rather than the Naval Observatory. Prior to this development, Espenak contributed to the available eclipse data by authoring his own *Fifty Year Canon of Solar Eclipses: 1986–2035.* This work was published as a NASA Reference Publication in July 1987 (NASA RP-1178). While somewhat hard to obtain today, it stands as one of the most prized and useful references for any serious eclipse chaser who wants to know "when and where is the next one of these?"

With the growing use of home personal computers, another even more convenient source of eclipse data has become available to the eclipse enthusiast. Companies including Zephyr Services of Pennsylvania, Carina Software in Northern California, and Future Trends Software in Texas have created programs for amateur and professional astronomers that can predict (often within 5 kilometers) the paths, contact times, and local circumstances of virtually any eclipse from 2000 years in the past into the dim future.

WEIGHING THE FACTORS

Once you know the west-to-east sweep of a specific eclipse path, the next major task is to consider thoroughly the possible places you might go to observe the event. You need to consider three principal factors: climate patterns, governmental and political stability, and the location's accessibility and infrastructure.

The most crucial element in choosing an eclipse site is meteorology. If the weather is unsatisfactory, the celestial show will be invisible. To get a good idea of the historical climatic patterns in the region, use publications like Anderson and Espenak's circulars, as well as all the data you can get from national or local meteorological offices in the country you're considering. However, even the best historical data will not guarantee good observing conditions on the day of the eclipse itself. To quote Isabel Martin Lewis, a pioneering female eclipse observer in the first half of the twentieth century:

> The weather has ever been the despotic ruler over the fates of eclipse expeditions. It is customary to get records whenever possible of weather conditions for a period of several years at points favorably located within the shadow path in order to minimize the chances of selecting stations where the weather promises to be unfavorable at the particular time of the year and day when the eclipse will occur. . . . The prediction of weather conditions, after all, is largely a lottery. . . . The astronomer has need to be a good sport or a philosopher, or both.[16]

CLIMATE AND WEATHER

Remember the oft-quoted meteorologist's credo: "Climate is what you expect, but in the end, weather is what you ultimately get."

"Fine," you say. "I need to obtain reliable, consistent, and recent historical weather information for this country. But I am not a meteorologist, nor do I have contacts in the country under consideration. How do I get this important data?"

The answer is simple: do some research. Find out if there is an amateur astronomy club in the capital city of the country. If so, what is its mailing address or telephone number? Does it have a fax machine? If no amateur club exists, get the most recent copy of the American Astronomical Society directory. At the rear of this publication, you can find an observatory or university astronomy department that will help you get the necessary meteorological data. Once you have the historical information, it's time to do some real analysis. Look at the cloud and sky coverage statistics over the past three to five years. In particular, see if you can determine any synoptic (large-scale) patterns to the timing, character, or intensity of the cloud coverage.

Remember that every place on Earth has clouds. The trick here is to do some crystal ball gazing and try to see the overall picture. Also keep in mind

Planning, for eclipse chasers, means thoroughly researching weather patterns, political conditions, and physical amenities in the viewing area well ahead of time. Those who find "the right spot" usually also find they're not alone. *Evelyn Talcott.*

that data from just one or two years in an area are not reliable. Shifts in the seasonal patterns happen. The recent recognition of the periodic El Niño effect in the southwestern Pacific Ocean has made meteorologists realize that historical weather patterns do not remain static—weird inversions of traditional climate do occur. You need to determine whether what you are analyzing is the typical pattern or an anomaly brought on by a relatively transient blip in the general scheme.

NARROWING YOUR CHOICES

When you have narrowed your search to two or three regions, you can begin unearthing the local infrastructure and accessibility of each specific area. Again, some detective work is called for. Go to a good book store and make your way to the travel section. Invest a few dollars and buy the most current, comprehensive guide books about the country or countries that seem to have the best weather prospects. Some of the better guides include those published by Lonely Planet books, the Fodor series, Fielding's Guides, and Insight Guide publications.

Read all you can about air service, hotel, and ground transportation options in the country. Are there regular, dependable flights scheduled from your home country and back? What is the range of hotel accommodations? Are there major automobile rental offices? If the topography is especially exotic, as in Bolivia, Indonesia, the Australian Outback, or Africa, can you rent four-wheel-drive vehicles?

What about the language(s) indigenous to the area? Do you speak the local lingo? If not, investigate the availability of a local, or "inbound" guide, as the travel industry calls them. Even if you can communicate with the locals, a guide is invaluable, especially if the place is not a usual tourist stop, or if its customs, laws, sanitation, or political stability are questionable.

If the destination you are considering is not in the headlines frequently (something that isn't all that bad), an outstanding source of recent data about the present political and tourist conditions is your government's state department or office of foreign affairs. A local tourist office is nice, but they usually paint a relatively positive picture of their nation. If the current ruler is on thin ice, or if the country's currency is about to go south in a big way, you are not likely to hear about it from the tourist board!

In the United States, a new service of the State Department can prove highly valuable. You can now call an automated information line in Washington D.C., and get hard copy of the latest tourist and travel data for over 190 nations. In addition, dedicated publications on passport acquisition, travel tips to certain regions, crisis situation procedures, and many other topics are available for just the cost of the phone call. To access this database, call (202) 647-3000 from a facsimile device with a touch-tone telephone handset. An automated menu will then guide you through the various options. You can get up to nine documents with a single phone call.

An important note here: Even though a specific area may pose many challenging problems in regard to tourist infrastructure, access, language, or transportation, do not rule it out immediately—particularly if the weather data indicate that the region offers a strong possibility for clear skies. With modern technology, it is possible to travel just about anywhere to observe an eclipse—even to the polar regions. The limiting factors are how badly you want to see the eclipse and your fiscal resources.

But again, one of your best bets is to depend on a reputable, savvy inbound tour operator or, better yet, a tour company experienced in operating eclipse trips. By letting someone else worry about the various arrangements, you are free to concentrate your energies on developing an observing program, refining your equipment list, and remembering crucial items for your journey.

What about your friendly local retail travel agency? Can't they make the arrangements and air reservations for you? All too often, the answer is no. Travel agencies provide a useful service to the public. They save travelers time and money and the frustration of navigating the intricacies of modern leisure travel. They seek the best deals in vacation programs and handle those small details that make a vacation successful. But when it comes to astronomical or scientific travel, retail travel agencies are way out of their pond. If you were to ask fifty retail agents where, when, and for how long the next total or annular eclipse will occur, you can be sure that fewer than five will answer correctly.

Furthermore, agents know nothing about astronomy, astronomical equipment, and the proper transportation of this equipment. They are unfamiliar with the hazards of observation during the partial phase of an eclipse and with sources of adequate solar filter materials. Anything having to do with choosing suitable camera film, exposure times, or focal lengths for capturing totality is alien to most travel agents. Add to the above the daunting task of finding out what countries, where in those countries, and how to gain access to the umbral path in time to see the event, and 99.99 percent of retail agents are overwhelmed with the technicality of arranging such travel.

CHOOSING A TOUR

What about a tour company, particularly one that is (or claims to be) experienced in eclipse excursions? As you would with any purchase, especially one of significant cost, be inquisitive. How long have they been in business? What is their success rate at seeing totality? If the firm's officers are not astronomers, do they have access to astronomers? Are they experienced eclipse observers? Do they operate other sorts of travel programs, or are they specialists in astronomical trips?

Have any of the company's trips failed to materialize, and if so, why (lack of participants, inability to get to the region of the eclipse, etc.)? Will they provide you with the names of past participants you can communicate with? Is the company a member of the U.S. Tour Operator's Association, American Society of Travel Agencies, or other groups? Do they belong to the Better Business Bureau? Most are dependable, helpful, and knowledgeable.

Be cautioned, though: Since the July 1991 total eclipse, many firms have sprung up claiming to know all there is to know about astronomical travel and about eclipses in particular. Many are simply retail agencies attempting to add a new wrinkle to their main business of writing airline tickets to New York, Omaha, and Seattle. Some may promise support and astronomical expertise but do not have any ready source of such information, let alone know what is involved in operating an eclipse expedition.

An excellent way to get unbiased information about tour firms is by reading eclipse preview articles in either *Astronomy* or *Sky and Telescope* magazine. Another option is to call on your local planetarium or observatory. Again, ask them if they know of a local firm that runs eclipse tours and how you can get in touch with them. If you or someone you know are a member of an amateur astronomy club, see if anyone in the group has heard of the tour you are thinking of joining.

Now that you have weighed all three factors that influence the choice of

Eclipse chasing doesn't involve the hardships it once did, but it will require some hard set-up work and you may have to endure extreme temperatures to get the view you want. *Barney Magrath.*

an eclipse trip, what comes next? If you have chosen to travel with a tour group, the rest is fairly easy. Your tour operator will probably ask for a deposit in good faith, to hold the place or places you want for the expedition. From then on, you will most likely be requested at periodic points to "ante up" with interim payments. This is not an unusual practice, even more than twelve months before the tour. Many local agents in far-off countries require earnest money to hold accommodations and secure other services so far in advance. It is especially true in cultures where such advance planning is normally unheard of.

On the other hand, if you have chosen to travel either alone or with a very small group of people, there remains more to do. You must secure three crucial things at this point—provided you can do so at all. They are, in order of importance, suitable hotel accommodations (for at least two days on either side of the eclipse date); reliable, assured air transportation to the closest major city to the eclipse path; and a suitable rental car or four-wheel drive vehicle. The remark "if you can do so at all" is not meant to be humorous or mawkish. One quickly learns that getting reservations far in advance of an eclipse is a daunting task.

A professional travel agent is helpful if you intend to do your own thing. However, several hard and fast rules must be invoked here (especially in light of the previous comments about the retail travel industry versus the needs of astronomical travel).

One: You dictate what you want to the agent. Don't let the agent steer you toward some canned tour package that includes elements you don't want. Agents make money through commissions. Other than this source of earnings, they are paid little or no salary. They are constantly bombarded by airlines, hotels, tourist bureaus, car rental companies, and other vendors to book people with a particular company, with the lure of additional money for this effort.

Two: Make sure the agent knows exactly where on Earth you're going and when you need to be in the staging city you have chosen. Tell them, again and again and again, the date, city, state, province or prefect, country, and—yes— the region of our planet that you intend to go to. This last part sounds moronic, but most retail agents will be unfamiliar with the area you want to see the eclipse from. Eclipses typically are best observed from somewhere off the beaten path. Most agents have difficulty understanding where you want to go, and why you would want to travel there.

Three: Get a final price for the entire package before any money changes hands. Then ask about trip cancellation insurance for you and your companions. If at all possible, charge your purchase on a major credit card (MasterCard, Visa, or American Express). Doing so protects you in case the hotel, air carrier, or travel agency goes out of business. Review the arrangements in detail and discuss them with your colleagues. If you see something you do not like or trust, tell the agent. Don't worry about being a pest. If the agent seems tired, bored, or unresponsive to your concerns and requests for further information, find a new travel agent.

CASING THE JOINT

This next step is optional, but if you have the time and money, it can go a long way toward insuring the success of your trip. This step is the site inspection. For some eclipses, it can be relatively simple. Such was the case for the total eclipse that occurred on July 11, 1991, and the annular solar eclipse of May 10, 1994. Both events happened in locations conveniently close to eclipse chasers—at least those in North and Central America.

However, with a couple of notable exceptions, the central eclipses for the remainder of this century will not be conducive to preliminary visits by anyone who is not seriously engaged in the operation of large tour groups. Visiting Bolivia, eastern Siberia, India, Vietnam, Indonesia, or the Australian Outback requires a significant investment of time and money. Nevertheless, performing a site inspection is a critical constituent of a good eclipse expedition and well worth the effort.

Why take the time to conduct such a trip? First, it affords you the chance

Cloud conditions often determine whether you'll need to travel far afield even if the eclipse is visible in your own back yard. Sometimes, as with this eclipse in Hawaii, the trip can be a vacation in itself. *Barney Magrath and Oluf Nielsen.*

to see the region well ahead of the event itself. Ideally, you should visit the region of the eclipse one year before the date of the eclipse, so that you can see an example of the possible weather awaiting you.

Second, a site inspection also allows you to view a number of places, so that you can make an informed choice about where you will ultimately observe the event. No map, travel guide, photo, or videotape can replace actually setting foot on the place where you intend to watch the eclipse.

Third, if you will be traveling with others, providing photographs of the areas you are considering will allow a consensus to emerge when the group gets together to discuss where to go. Also, by staying in hotels you intend to use, tasting the local food, and generally absorbing everything you can about the setting, you are light-years ahead of others who will have to do all this analysis after arriving only days before the eclipse.

Another advantage to site inspections is that you can select an observing area well ahead of others who are sure to follow. After you have visited the viewing area, you will already know how long it takes to travel there, the condition of the roads, the landmarks en route to the site, and where to set up your equipment.

If you intend to employ a local inbound guide or tour company, a site inspection is the prime opportunity to establish rapport with the people you will be depending on to get you into the umbral path. If you come prepared with detailed maps, eclipse path plots, and other data, it will make the tour company's job that much easier. You can show them precisely what you want and expect from them on the day of the eclipse. When you can visually show tour personnel where you wish to see the event, they can quickly give you both the positive and negative aspects of the region you are interested in.

EQUIPMENT

For most of the remaining eclipses in the twentieth century, travel light. Most of the countries the eclipses will be passing through are far from North America, Europe, and Australasia. Thus, you won't want to be weighted down with armfuls of luggage and peripheral equipment. As you will see in Chapter 6, you can photograph a solar eclipse effectively with relatively simple, lightweight equipment. You don't have to tote massive tube assemblies and equatorial mounts or drives to some far-off land.

Before actually packing for the trip, keep several things in mind:

First, if you are part of a group, will the luggage be checked in as a group or individually? The former arrangement gives you some leeway, particularly if you have more than the normal amount of equipment or bags. With group-checked baggage, airlines usually take the total baggage weight and divide it by the number of people traveling together; so you can sometimes fudge a little and exceed your limit slightly, if necessary.

Second, if you are traveling alone or with an independent group, find out from the airline the maximum allowable weight and size of checked luggage. Also inquire about additional weight or size charges for extra bags or boxes a passenger may wish to take. Most international carriers charge a substantial sum for going over their limits—this will do more to pare down your list than any other deterrent.

Third, remove the optics from your telescope or tube assembly before checking your luggage, if possible. Better yet, take the optics aboard as

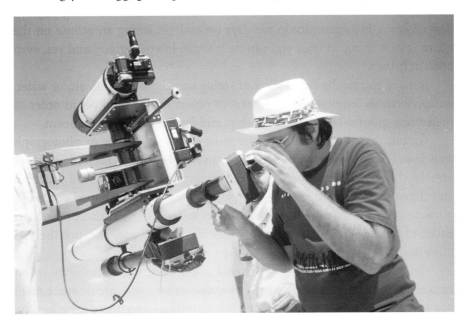

Traveling abroad with sophisticated equipment means taking such extra precautions as hand-carrying fragile optics and registering cameras and telescopes with U.S. Customs before you leave to establish its origination. *Jeff Schroeder.*

carry-on baggage to ensure safe handling. This forces you to keep the aperture of any optics to a reasonable diameter, and you will leave that 20-inch Newtonian telescope at home, where it really belongs.

Fourth, pre-register any significant camera or optical equipment you own with your customs department prior to your departure. In countries like the United States, this is particularly important. Pre-registering your hardware will unimpeachably establish its pedigree and origination, and thus keep you from possibly having to pay import or duty fees upon your return home. Eclipse chasers all have tales of woe about having to explain to a customs officer that those Fujinon binoculars or that Nikon camera body and lens were purchased at home, not in the country they were returning from.

By Land, Sea, or Air

We live in a marvelous era. We can access any corner of the Earth by air in a dozen or so hours of travel. Gone are the days of arduous, week-long ocean voyages on wind-driven vessels, eating tasteless food, risking physical danger, being incommunicado for days on end. If seeing an eclipse on the ocean is your thing, at least you can do it today in style, grace, and yes, even opulent luxury.

If asked to rate the desirability of seeing an eclipse by land, air, or water, our preferences following twenty years of observation would be, in order of preference: ground-based, aboard a ship, and from an airborne platform.

Ground-based observation is the most stable way to see or photograph an eclipse (except when there is seismic activity, of course). Even in calm seas any vessel always has motion due to ocean currents or winds. And watching an eclipse from a pressurized, cramped, isolated tubular-shaped container hurtling over 500 miles an hour, seven miles above Earth's surface, makes the entire experience akin to watching a movie of an eclipse, and not the real thing itself.

On the other hand, ships provide almost limitless mobility when required to find an area of cloud-free sky, and jet airplanes are immune to all but the most complex and mighty storm systems—virtually assuring the eclipse chaser of seeing totality.

There are no inviolate rules in picking eclipse viewing sites. Certainly, careful collection and review of historical data and current political and infrastructure conditions are critical, but a well-seasoned set of gut feelings can go a long way toward making the difference between eclipse-day success and bitter disappointment.

A dramatic photo like this can wipe out the jet lag of a trip halfway around the world and keep you coming back for more. *Ernest Piini.*

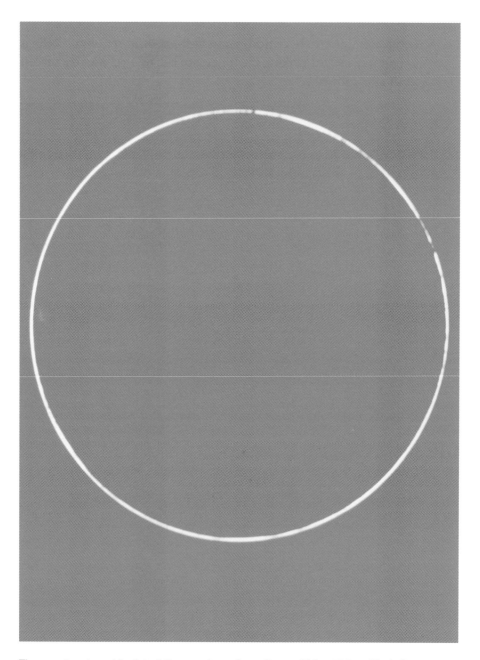

The spectacular midpoint of the annular solar eclipse of May 1984. *Mark Coco.*

5

VIEWING AN ECLIPSE

Once you've made the decision of where to view an eclipse, your next decision is how to view the eclipse: by the naked eye, with binoculars, or through a telescope. There is no one best approach—in fact, we recommend pursuing all three methods because that will let you view the full range of eclipse phenomena.

EYE SAFETY

Your first concern must be to protect your eyesight. Just a brief glance at the Sun during the partial phases of an eclipse can damage your vision and even cause blindness. See Chapter 7 to learn how to safeguard your eyes. The good news is that you need protection only during the partial phases of the eclipse. During totality you can view the Sun safely, even with optical aid.

Your eyes are by far the best instrument for getting an overall view of the eclipse. You can't beat them for viewing the twilight colors that ring the horizon during totality, watching the Moon's approaching shadow, or taking in your surroundings as the eclipse progresses. Perhaps most importantly, the naked eye delivers a great view of the eclipsed Sun hanging in a darkened sky.

Binoculars magnify what you see and provide more detail at the expense of narrowing your field of view. They are good for viewing the partial phases of an eclipse and spotting bright stars and planets that come out as darkness approaches. You can use them to examine the size and shape of the Sun's corona during totality.

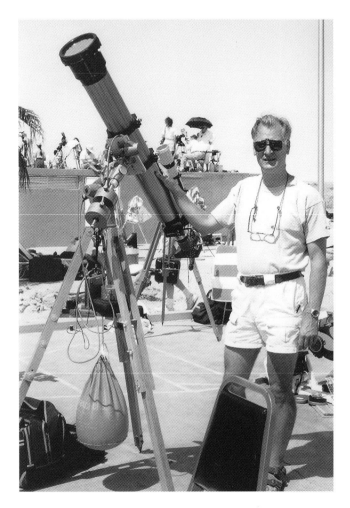

Sunglasses won't protect your eyes during partial phases of an eclipse, and only an over-the-aperture solar filter allows viewing or photographing these phases through a telescope. *Dan Good.*

If you want a closeup view of the Sun during an eclipse, however, you'll need a telescope. Only a telescope provides enough power to show you fine detail on the Sun during the partial and total phases. A scope will reveal sunspots on the face of the partially eclipsed Sun, the ruddy hue of the Sun's chromosphere just before totality commences, and details in solar prominences and the inner corona during mid-totality.

But using a telescope to observe the Sun is far different from using a telescope for observing at night. The most important difference, of course, is that you must take safety precautions. Either use a good over-the-aperture solar filter or project the image of the Sun, as described in Chapter 7. And be sure to keep your finder-scope capped, so that no one looks through it accidentally.

To view the Sun with your telescope, you first have to find it. How can you do that with the finder capped? The easiest way is to make the shadow that is cast on the ground by the telescope's tube as small as possible. Turn the knobs that control where the telescope points until the shadow shrinks to a minimum. Your telescope should then be aimed right at the Sun. The Sun's image should appear in the eyepiece if you're viewing with a filter or on the screen of a projection setup.

AS THE SHADOW NEARS

A typical total eclipse lasts two to three hours from first to last contact. At first this might seem like plenty of time to see and do everything you want to, but you'll quickly find that's not the case. Time has a strange way of seeming compressed during an eclipse; the fleeting moments pass quickly. To compound the problem, most of the action and excitement occur during the few moments right around totality. If you intend to view the eclipse with a telescope or to photograph it, plan to arrive at your observing site at least an hour or two before first contact. That will give you plenty of time to set up your equipment and to be sure it's operating properly. After all, you don't want to find out your camera's not working just as totality begins.

The eclipse officially gets under way at first contact, when the edge of the Moon's disk first encroaches on the Sun. No one can actually see first contact because it marks the instant when the disks of the two main players lie tangent to one another. It's not until a moment later, when the Moon's disk covers a perceptible part of the Sun's disk, that you can see anything.

Everyone who watches an eclipse eagerly awaits first contact. Although predictions made years before foretell the exact moment, people like to see it for themselves. If you're part of a group, you can count on a contest to see who first spots the telltale nick in the limb of the Sun. When the winner cries out—"There it is, first contact!"—everyone else hurries to their instruments to confirm the sighting.

The next hour or so is one of breathless anticipation. If you don't look at the Sun occasionally to see that the Moon is making steady progress, you might not be aware that an eclipse is taking place. The signs are subtle—the sky grows darker, but the pace of change is slow and difficult to discern.

After the Sun is about half-covered by the Moon, a perceptive eye starts to notice changes. Shadows become sharper as the source of light in the sky becomes smaller. (The sharpness of a shadow depends on the apparent size of the light source. The same effect makes shadows cast by a large, coated incandescent bulb or a long fluorescent fixture appear softer than that from a naked filament.)

By the time the Moon covers about 80 percent of the Sun's disk, the sky has grown noticeably darker. (Although to a person on the street not paying attention to the ongoing eclipse, the changes would be easy to miss.) This is a good time to shift your gaze briefly from the sky to the ground. Look closely at the sunlight filtering through the leaves of nearby trees and bushes and you'll see hundreds, if not thousands, of tiny images of the crescent Sun. The openings between the leaves serve as pinholes, and each projects an image of the eclipsed Sun.

VIEWING THE STARS AND PLANETS

As the Sun wanes to a thin crescent, about ten minutes before totality begins, start scanning the sky for bright planets and stars. At this stage you may spy only a few of the brightest ones, but as totality approaches more will come into view. During totality, you can expect to see objects as faint as third magnitude.

As the partial phase of the July 1991 eclipse progressed, observers in Baja California viewed the Sun's shrinking crescent through telescopes and watched as their own shadows grew sharper. *Evelyn Talcott.*

Which stars and planets will appear depends on the time of year the eclipse takes place, where in the sky it occurs, and where the bright planets lie at eclipse time. For instance, during the total eclipse of July 11, 1991, viewers in Hawaii who had clear skies saw many of the bright stars of the winter sky—most notably Sirius, Betelgeuse, Rigel, and Capella. Those in Baja California, on the other hand, got to see the same stars, as well as the bright planets Venus, Jupiter, Mars, and Mercury. At the time of the eclipse, those four planets all lay east of the Sun. Hawaiian observers missed seeing the planets because the eclipse occurred low in the eastern sky, and the planets had not yet risen. From Baja California, the eclipse occurred nearly overhead, so the planets were all high in the sky.

To determine which stars should be visible during an eclipse, you need to know what time of year the eclipse will occur and what sector of the sky the Sun will be in at eclipse time. The Sun takes a full year to travel once around the sky relative to the background stars, and it blots out most of them in the daytime sky. The stars that become visible during an eclipse are those stars that appear at night when the Sun is halfway around the sky from its position during the eclipse. That's why the winter stars were prominent during the July 1991 eclipse. The November 1994 total eclipse occurs chiefly in South America, where it will then be late spring, so the stars of the southern autumn sky will become visible behind the darkened Sun.

Next take into account where in the sky the Sun will be. If the Sun is high in the sky at eclipse time, as it was in Baja in 1991, then you'll see the sky as it appears around midnight six months hence. If the Sun is just rising or setting, then the star background will resemble the sky seen at sunset or sunrise, respectively, six months down the road.

Knowing which planets may be visible is more difficult because these objects move relative to the background stars. One way to determine the planetary alignment is to find out which planets are visible, and at what time of night, in the days leading up to the eclipse. Planets that are visible in the evening twilight glow will appear to the east of the Sun during the eclipse, while those in the dawn sky will appear west of the Sun. Those planets near opposition—opposite the Sun in our sky—won't be visible because they are below the horizon. The diagrams in Appendix B show the sky surrounding the eclipsed Sun for each of the total eclipses during the rest of the decade.

TOTALITY APPROACHES

About four to five minutes before second contact you can begin to look for shadow bands. They are a mysterious, subtle pattern of light and dark bands that ripple across the ground shortly before and after totality. They appear much like the pattern of ripples you see at the bottom of a swimming pool on a sunny day. They travel at about the speed of a jogger, some 5 to 10 miles per hour.

As testimony to the subtlety of shadow bands, many eclipse veterans have never seen them. Surprisingly, many even choose not to look for them—arguing that there are more than enough exciting phenomena taking place in

Some eclipse observers watch the ground for phenomena like the multiple eclipse images seen here shining through the leaves of a tree. Others watch for shadow bands on the ground. *Jay Friedland.*

the final moments before totality. On the other hand, some eclipse chasers consider shadow bands to be among the highlights of an eclipse. Perhaps the best advice we can offer is that if you are viewing your first total eclipse, don't let the pursuit of shadow bands cause you to miss any of the other interesting effects. But if you have seen an eclipse before, consider spending a few minutes to look for them. Another possibility is to try taping shadow bands with a tripod-mounted video camera. That way you won't be distracted from the other goings-on but can still view them later.

The best place to spot shadow bands is on a smooth white surface. Try staking a large white sheet to the ground. Make it as taut as possible—you don't want the wind creating ripples in the sheet that could mimic the real shadow bands. Alternatives include a large piece of white cardboard or a sheet of plywood that has been painted white.

Astronomers still debate the cause of shadow bands. Some think they are interference patterns. As the Sun shrinks to a slender crescent, the light waves passing through pockets of slightly different densities in Earth's atmosphere interfere with one another. Where the interfering waves add together, you get light bands; where the waves cancel each other, you get dark bands. Others think shadow bands are created by turbulence in the atmosphere. In this case, shadow bands not only look like the ripples seen at the bottom of a swimming pool on a sunny day, they also form in a similiar way.

Once you've looked for the shadow bands, turn your full attention back to the Sun. As the solar crescent dwindles to a tiny sliver, eventually it begins to break up. Sunlight no longer forms a complete arc around the limb of the Moon. Instead mountains along the Moon's limb poke up through the ever-thinning crescent. As the Moon continues its relentless march, more mountains cut off the Sun's light, and the only places sunlight passes through are valleys along the lunar limb.

This string of brilliant points of sunlight arrayed along the Moon's advancing edge is known as Baily's beads, after Francis Baily, the 19th-century English astronomer who first described them in detail. Baily's beads appear during both total and annular eclipses. They never look exactly alike. Not only does the Moon's limb shift slightly from one eclipse to another, but the geometry of Sun, Moon, and Earth changes, and the speeds of the three bodies vary.

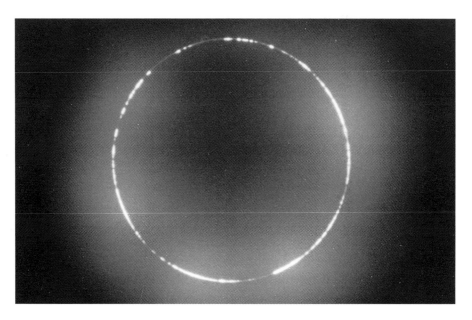

Top: Baily's beads, which are caused by sunlight passing through lunar valleys. *Huguette Guertin.*
Bottom: The diamond ring, just after third contact. *Doug Murdock.*

As Baily's beads dwindle in number, steal another few seconds from viewing the Sun and look toward the western horizon. About fifteen seconds before totality you'll see the Moon's shadow descending on you like a huge wall of darkness. Traveling at supersonic speed, the looming shadow awes almost everyone who sees it as it announces the imminent arrival of totality.

With just a few seconds to go, the jewel-like Baily's beads shrink to a final solitaire, known as a diamond ring. This last bit of sunlight streams through the deepest gap on the Moon's limb, providing the disappearing Sun with one last moment of glory before totality reigns.

SECOND CONTACT

As soon as the diamond ring is extinguished, remove the solar filters from your telescopes, binoculars, and cameras. For the duration of totality, you don't have to worry about protecting your eyes or equipment from the Sun's radiation. The harmful surface of the Sun, the photosphere, is hidden safely behind the Moon.

The next feature you'll see pops into view along the Moon's advancing limb about five seconds after totality begins. It is the chromosphere, the thin layer of the Sun's atmosphere just above the brilliant photosphere. The chromosphere appears a vivid red, courtesy of hydrogen atoms emitting radiation at a specific wavelength in the red part of the spectrum. But look quickly—the chromosphere disappears within a few seconds; it is only about 1,000 to 2,000 miles thick, so the Moon soon covers it up.

Once the chromosphere is hidden, you're in the glory of totality. The most spectacular sight is the Sun's corona, a gauzy, pearly white light that typically stretches two or three times the diameter of the uneclipsed Sun. Look for delicate details throughout the corona: loops, swirls, and streamers commonly appear.

Although the corona comes into view as soon as totality begins, it will take a moment for your eyes to grow accustomed to the sudden darkness and to appreciate all the glorious detail. You may be surprised at how bright the corona appears—it's about half as bright as the Full Moon. You don't normally see it because the Sun's photosphere drowns out the corona's light.

The corona forms the outermost region of the Sun's atmosphere, extending millions of miles above the surface. It's also the hottest part of the atmosphere, glowing at temperatures of a million degrees. It would appear even brighter than it does if the gas that makes up the corona weren't so rarefied. The size and shape of the corona vary from one eclipse to another. When the Sun is active—when there are lots of sunspots and other forms of magnetic activity—the corona tends to appear round and symmetric. When solar activity is low, the corona appears asymmetric and has many streamers.

Although it might be tempting, don't spend all of totality looking at the corona. Other things going on are well worth a look. The other major feature to watch for on the Sun is prominences—fiery red tongues of gas that arc above the chromosphere into the corona. They are cooler and denser than the rest of the corona's gas, and their shapes are controlled by the Sun's magnetic field. Some prominences are usually visible throughout totality, although near the middle of some long eclipses the Moon covers enough of the Sun's inner corona to hide them all from view.

It's also worth tearing yourself away from the Sun during totality to take in your surroundings. Scan the sky to see how many planets and stars you can spot. And look at the twilight colors that ring the entire horizon. These colors come from sunlight just outside the area of totality.

WRAPPING IT UP

Almost as soon as totality begins, it seems to come to a close. Third contact—when the first glint of sunlight from the photosphere appears and drowns out the corona—signifies the end of totality and the time to replace all the filters on your equipment.

But the eclipse is far from over. Everything you saw between first and second contact now plays itself out in reverse. And with the anticipation of totality no longer hanging over you, you may find yourself better able to appreciate these phenomena.

First you'll see the diamond ring effect, coming into view on the side of the Moon opposite where the first one was. Then you'll see Baily's beads and the elusive shadow bands. Watch too as the Moon's shadow speeds off to the east, and imagine the other lucky souls out there who no doubt wait for their view of totality.

CHECKLIST OF ECLIPSE EVENTS

Before Totality
- First contact
- Moon covers sunspots
- Shadows on the ground sharpen
- Projected crescents
- Sky darkens
- Temperature drops
- Shadow bands
- Baily's beads
- Moon's shadow approaches
- Diamond ring
- Chromosphere

During Totality
- Prominences
- Coronal structure
- 360° twilight
- Planets and bright stars

After Totality
- Chromosphere
- Diamond ring
- Moon's shadow recedes
- Baily's beads
- Shadow bands
- Temperature rises
- Sky brightens
- Projected crescents
- Shadows more muted
- Moon uncovers sunspots
- Last contact

A partial solar eclipse projected onto a sidewalk by the branches of a palm tree makes an interesting subject for photography. *Daniel P. Joyce and James Carroll.*

6

CAPTURING AN ECLIPSE ON FILM

Nothing quite matches seeing a solar eclipse with your own eyes. Whether it's the Moon slowly but steadily pushing its way across the face of the Sun, the bright ring of sunlight that marks the culmination of an annular eclipse, or the ruddy prominences and pearly white corona that highlight totality, the image of the eclipse will remain in your mind for a lifetime. Even so, most people who view an eclipse also want to record it on film or videotape. Many think eclipse photography is an art only a select few can master. Not so—capturing stunning images of the eclipsed Sun is not that difficult. In fact, in many ways it is easier than other forms of celestial photography.

All you really need to capture an eclipse is a camera and a tripod. Of course, you can go the expensive route if you want to, using fancy gear with all the latest bells and whistles. But as long as you focus precisely, use fine-grained film, and have a steady mount, you can capture lasting memories with the simplest of equipment.

CAMERA ON TRIPOD SHOTS

What kind of photos can you hope to get with a typical 35mm single-lens reflex camera mounted on a sturdy tripod? With a normal 50mm lens, the Sun appears just under .5 mm across on the film. You won't capture any great prominence detail or fill the frame with the glorious corona. But you

Video cameras, properly protected by filters, can provide a stunning record of an eclipse, including the sounds of other observers. *Ernest Piini.*

will get wonderful scenic shots with an eclipsed Sun hanging in the background. Such a picture probably captures the feeling of being at the eclipse better than any other type of photo. If you can record totality in your shot, the eclipsed Sun will look like a hole in the sky surrounded by the opalescent corona. But even a partial eclipse can look spectacular if captured with a scenic foreground.

The key to getting a great photo is to plan the shot well ahead of time. Know where the Sun will be when it is in eclipse. If it will be near the horizon, scout out the perfect foreground to create the greatest pictorial impact.

A standard 50mm lens on a 35mm camera captures a field about 40° wide by 27° tall. (If you rotate the camera 90°, you'll get a field 27° by 40° instead.) That gives you some leeway for capturing the eclipsed Sun along with a foreground scene, but it may not be enough. If it isn't try using a wide-angle lens. A 35mm lens will give you a field 54° by 38°, while a 28mm lens yields a field 73° by 50°.

Even if the eclipse takes place near the zenith, however, you're not out of luck. Particularly during a total eclipse, consider using a fisheye lens to capture the entire sky along with the horizon. Simply aim the camera straight up and you'll record the surreal 360° twilight glow that rings the horizon. You can mimic the effect of a fisheye lens by laying a shiny hubcap directly on the ground and photographing it with a normal lens.

There's one other camera-on-tripod shot worth taking during totality. That's one designed to record any planets or bright stars that pop into view in the suddenly darkened sky. To capture the planets and stars along with the Sun, you need to know where they will be and what size lens will encompass the area. Because the solar corona outshines most of the planets and stars, the corona has to be overexposed if you want to see the stars. Usually a 5- to 10-second exposure on moderate-speed film works well.

SOLAR CLOSEUPS

If fiery red prominences or coronal streamers are your cup of tea, you'll need a lens with a focal length much longer than 50mm. The size of the image on the film increases proportionally to the focal length of the lens, so a 100mm telephoto lens gives you a solar image twice as large as a 50mm lens does, or about .9 mm across. A 200mm lens increases that by another factor of two. You can always figure out the size of the Sun on the film frame, in millimeters, by dividing the focal length of the lens by 110.

To magnify the Sun enough to see detail clearly, use a focal length between 750mm and 2,000mm (corresponding to an image size of from 7 mm to 18 mm). One way to achieve these long focal lengths is to use a telephoto lens with a 2x or 3x teleconverter to effectively double or triple the focal length of the lens. Keep in mind that teleconverters rarely have the optical quality of a good lens, so if you have the choice of using a longer focal length lens instead of a teleconverter, do so.

But probably the best way to get a long focal length is to use a telescope as your camera's lens. You don't need a massive scope for your system. A 3- or 4-inch telescope not only works well but can be transported easily if you are traveling to your eclipse site. The camera body attaches directly to the focuser of the telescope with a camera adapter.

Because the long focal length of a telephoto lens or a telescope magnifies the Sun so much, the slightest motion of the camera or lens can noticeably

Using longer focal length lenses or a telescope as your camera lens can yield close-ups of Baily's beads and solar prominences. *Ernest Piini.*

blur the image. That's why a sturdy tripod or telescope mounting may be the best investment you can make in eclipse photography.

A sturdy mount by itself doesn't guarantee rock-steady images, however. You need to seek out and eliminate other sources of vibration that can turn your tack-sharp photos into fuzzy blurs. If you're observing with a group of people, try to set up well away from the crowd. You don't want someone to bump into your equipment accidentally, and you don't want any overexcited observers to start jumping up and down next to you right during the precious moments of totality.

Not all sources of vibration are external, however—the camera itself can contribute to blurry images. Every time the shutter is released, a mirror inside the camera flips out of the way to allow light to reach the film. But the slap of the mirror inside the body shakes the camera and can blur the images. If your camera allows it, lock the mirror out of the way.

Even pressing the shutter can induce a small vibration that shows up in the final photo. At the very least, use a cable release or, even better, the camera's self-timer (if it has one) to trigger the shutter.

DRIVING YOUR TELESCOPE

After concentrating on tiny motions inside the camera, it's time to turn to what at first seems the largest motion of all—that of the Sun across the sky. It appears to move, of course, because Earth rotates beneath us. If you simply pointed a camera at the sky and opened the shutter, the Sun would trail across the frame. The question is, how long does an exposure have to be before that motion shows up on film?

Although the answer depends on the focal length of your system, most eclipse photos are luckily shorter than the critical value. You can calculate the maximum exposure that won't show trailing by dividing the focal length of the lens or telescope in millimeters into the number 250. Thus with a 250mm lens, you would start to see trailing after 1 second. Even with a 1,000mm-focal-length telescope, you could expose for 1/4 second and still get razor-sharp photos. Because most eclipse photos, except those trying to capture the faint outer tendrils of the corona during totality, require exposures shorter than this, there's no need to use a clock drive to compensate for Earth's motion.

That doesn't mean you shouldn't use a clock drive, however. The big advantage it gives you is that the Sun will stay centered in your camera's field of view. Our star moves across the sky about 1°—two solar diameters—every four minutes. When viewed at high magnification through a telescope it rapidly slips out of the field of view. By tracking the Sun, a clock drive helps you keep it nicely framed.

Keep in mind that for the clock drive to perform well you need an equatorial mount that has been polar aligned properly. The best way to do this is to go to your eclipse site the night before and align on the celestial pole. If you're traveling from the Northern to Southern Hemisphere or vice versa, make sure you can polar align in the unfamiliar sky and that the drive's direction can be reversed to operate in the opposite hemisphere.

BEST FILMS AND EXPOSURES

Choosing the right film for capturing an eclipse is a tradeoff between film speed and grain size. The "faster" a film is, the quicker it records light, so you can keep exposures short and minimize motion in the image. But faster films have more grain than slower films, so they don't record detail as well.

Fortunately eclipses, even total ones, are bright enough that you can use a slow film and still keep exposure times fairly short. A film with an ISO rating of 100 or so will capture sharp detail and retain it when enlarged. Good print films include Ektar 125, Kodacolor 100, and Fujicolor 100. For slide film, try Kodachrome 64, Ektachrome 100, or Fujichrome 100. You can also get nice results with ISO 200 film, though enlargements show a little grain.

Faster films show more noticeable grain, but they have one key advantage. If you're viewing an eclipse from a cruise ship, the constant rolling motion can play havoc with your photography. An ISO 400 film will help keep exposures short and image motion down. Your best bet on open water is to anticipate when the ship's rolling motion will change direction, which is when it will be moving most slowly.

The choice of prints or slides comes down more to personal preference than to any inherent advantage of one over the other. Print film has a slightly larger "exposure latitude," meaning that your shutter speed or f/stop can be a bit off yet still give nice results. Slide film, on the other hand, gets processed in one step so all the detail you record appears on the transparency.

SHOOTING THE ECLIPSE

Photographing eclipses with a telephoto lens or through a telescope can be divided into two distinct parts: recording the partial phases and capturing totality. For the partial phases—for the purposes of photography, they include the ring of sunlight visible during an annular eclipse—you need to use a filter to protect your camera from the Sun's heat. During totality, the filters come off and you shoot the Sun directly.

The best filters for the partial phases are aluminized Mylar or metal-on-glass types that fit over the aperture of the telescope or the front of the telephoto lens. You have a little more latitude with photographic filters than you do with filters for viewing the Sun directly because the eye is sensitive to ultraviolet and infrared radiation, while film is not (see Chapter 7).

You can also use a #12 to #14 arc-welder's glass as a filter material. Optically these filters leave a little to be desired and they turn the Sun a disconcerting shade of green, but they will give you reasonable images. Finally, you can use a 5.0 neutral density photographic filter. (However, NEVER look directly at the Sun with such a filter because it transmits dangerous levels of infrared radiation.) When using a neutral density filter, shade the camera between exposures to avoid overheating either the camera or the film.

For the partial phases, exposures will probably run about $1/125$ second at f/8 with ISO 100 film, but the exact exposure depends on the filter and the sky conditions that day. Your best bet for getting perfect photos is to make a test run before eclipse day. The surface brightness of any part of the Sun remains constant even when another part of the Sun is eclipsed. So you can go out a week before the eclipse and find out what exposure will work for your setup.

With your filter in place, expose a roll of film using a wide range of exposure times and different f/stops (if you're practicing with a telephoto lens). Get the film developed and see which combination worked best. On eclipse day, shoot the partial phases using the same exposure and f/stop. To guarantee getting some good shots even if the sky conditions are different, bracket your exposures by one or two f/stops or shutter speeds.

Keep using the same f/stops and exposures until only a thin crescent remains. The limb of the Sun appears slightly darker than the rest of the disk,

Viewing the eclipse from shipboard gives you mobility to get to a clear viewing area if you encounter cloudiness on eclipse day. This advantage may be offset by rough seas, however, if you plan to photograph. *Spencer R. Rackley IV.*

an effect that arises because we see higher and cooler layers of the photosphere when we look toward the limb. To counteract this limb-darkening effect, increase the exposures by one or two stops near second contact.

Contrary to popular belief, exposure times during totality aren't all that important—almost any length exposure will give you a good shot. A short exposure captures the reddish prominences, an intermediate exposure reveals the inner corona, and a long exposure brings out the outer corona. The only thing you need to make sure of is to take off the filter during totality or you won't record anything at all.

Follow a simple plan for your total eclipse shots. About 10 to 15 minutes before totality, load your camera with a fresh roll of 36-exposure film. You won't have time to change it during totality, so do it beforehand,

A convex, or "fish eye," lens gives you a photograph of the eclipsed Sun in a context that includes 360° of twilight around the horizon as well. *Spencer R. Rackley IV.*

even if you still have unexposed frames left on the current roll. After totality begins and you've taken the filter off, manually focus your camera as accurately as you can. A bright prominence or the dark limb of the Moon makes a good focusing target. The time you spend focusing will pay off handsomely in crisp images.

With a telephoto lens, you'll want to set the f/stop to f/4 to f/5.6. To capture the fleeting Baily's beads or the diamond ring, shoot at speeds of $\frac{1}{1000}$ to $\frac{1}{125}$ second. Once the diamond ring disappears, keep the exposures short ($\frac{1}{500}$ to $\frac{1}{60}$ second) to get good photos of prominences. Then gradually lengthen the exposures up to a maximum of 1 or 2 seconds to grab the corona. Of course, the length of totality will dictate what you can accomplish.

Follow a similar routine with a telescope, though the exposures will vary

For dramatic effect, a photograph which includes an eclipsed Sun, the sea, and a horizon is hard to top. *Jorge Soto.*

depending on the scope's focal ratio. (This number is probably marked on the telescope's tube. If it isn't, divide the scope's focal length by the diameter of its main lens or mirror measured in the same units.) With a fast scope (f/4 to f/8), try the same times recommended for the telephoto lens above. For an f/10 scope, you'll need to increase the exposures by about one stop, and for an f/16 scope, you'll want to go up an additional stop.

A WIDE-ANGLE VIEW

If you intend to shoot wide-angle or scenic shots of the eclipse instead of solar closeups, try a broad range of exposures. Open up the lens to f/2.8 to f/4 and expose for times ranging from about ⅛ second up to 10 seconds. The shorter exposures should more accurately reflect your visual impression of the eclipse, while the longer exposures will reveal fainter stars and landscape details you may not have seen.

If you're feeling ambitious, you might consider taking a different kind of wide-angle shot. By using a camera capable of taking multiple exposures, you can capture the entire eclipse on a single frame of film. The key to making this work is knowing exactly where the Sun will be during the eclipse. Head out to your observing site a day or two before the event and determine precisely where the Sun will be. That should enable you to keep the Sun nicely framed in the field of view on eclipse day.

Beginning a few minutes after the partial eclipse starts, make one exposure every ten minutes. Keep the interval between each shot precisely the same so that the images of the Sun will be evenly spaced across the frame. Keep the filter on the camera for all the exposures unless you're in the path of totality. If that's the case, take the filter off for one long exposure of the totally eclipsed Sun.

VIDEOGRAPHY

Many observers bring camcorders along to make movies of their eclipse experience. Surprisingly, camcorders are well designed for the rigors of eclipse recording. Of course, you will need a solar filter to protect the light-sensitive CCD during the partial phases of the eclipse, but during totality you should take it off.

As in regular photography, you'll want to mount the camcorder on a tripod so that the image doesn't move around. The maximum zoom on many camcorders magnifies the Sun enough to yield an image over an inch in diameter on a TV screen. And you can double this (or more) by adding a converter lens. One of the highlights of taping an eclipse is the soundtrack. The voices, the shouts, the noise—all help to bring back the experience more than simple photos do.

Perhaps the most important thing to remember about capturing an eclipse on film or videotape is not to forget that you want to see totality with your own two eyes. More than one eclipse chaser has gotten so immersed in the photography that they failed to look at the eclipse until totality was over. Build time into your routine to stop and look around, to remind yourself why you came to see this event in the first place.

Totality—the only stage of a solar eclipse that's safe to watch without eye protection. *Ronald Royer.*

7

SAFETY FIRST

Ask people what their most precious sense is, and they're likely to say their eyesight. The majesty of an eclipse needs to be experienced with your eyes, but its beauty also brings danger. The problem stems from the brightness of the partially eclipsed Sun. It takes only a tiny sliver of the Sun's disk viewed for a few seconds to injure your eyes, perhaps permanently. Fortunately, you can safely view the totally eclipsed Sun without protection.

THE DANGERS

Who among us hasn't glanced briefly at the Sun sometime in our lives? But we looked away just as quickly—the glare was too great. Two factors make viewing the partial phases of an eclipse particularly dangerous: First, the eclipse makes us want to look at the Sun; and second, since the total brightness of the Sun is diminished, it is easy to gaze at the Sun for long periods.

The human eye functions like a lens, focusing incoming light on the retina. If you've ever taken a magnifying glass and focused the Sun's light on a piece of paper, you've seen how easily sunlight can burn. The same thing happens when you look at the Sun, except that what burns is the delicate tissue of the retina. It happens quickly; the retina of an unprotected eye can burn in just thirty seconds, even with no optical aid. Looking at the Sun through binoculars or a telescope can blind you in a fraction of a second. There's no warning this damage is taking place: The retina has no pain receptors, so any damage happens painlessly; the visual effects don't appear until hours later.

A thermal lesion is the worst kind of damage. It occurs when sunlight raises the temperature of the retinal cells some 40° Fahrenheit, to about 140° F. The heat destroys the cells of the fovea centralis, the central part of the retina that provides the sharp views at the center of the field of vision. With this area destroyed, you can no longer read, drive a car, or perform other activities that require sharp central vision. Thermal lesions arise from long-wavelength visible radiation and infrared radiation.

The second and more common kind of damage is a photochemical lesion. In this case the temperature of the retinal cells needs to rise only 4° F. The damage is not only less severe but also in some cases only temporary. Unlike thermal lesions, photochemical ones develop when the retina absorbs short-wavelength visible light, usually blue or violet.

Despite the dangers of viewing the Sun directly without protection, many people still fall victim. The great Italian astronomer and physicist Galileo Galilei, who expanded our horizons with his newly invented telescope, became totally blind late in life, quite possibly because he spent too much time observing the Sun with his small refractor. But victims of the Sun's light aren't confined to the dim past. On March 7, 1970, during the last total eclipse visible from the eastern United States, nearly 150 people suffered some eye damage during the partial phases. Half of them never regained their full eyesight.

SAFE VIEWING METHODS

To avoid these hazards, you must take precautions whenever you view the partially eclipsed Sun. If you plan to view the eclipse directly, you'll need a high-quality filter that blocks not only most of the Sun's visible light but also its infrared radiation. A filter is necessary whether you plan to view with the naked eye or through binoculars or a telescope. (When viewing an eclipse through a telescope, remember always to keep the lens cap on the front of your finderscope. Curious people, particularly children, are apt to look through the finderscope without thinking, not realizing the extreme danger.)

Three types of filter material work well. An inexpensive but perfectly safe filter is a #14 arc-welder's glass. This dark green glass blocks all the Sun's infrared and ultraviolet radiation and slashes the visible light by a factor of

Nature's own solar projection: multiple images of a partially eclipsed Sun shine through palm leaves onto a sidewalk in Baja California. A simple homemade projection box can provide observers with a safe image of the eclipse. *David M. Beraru.*

about 370,000. Be sure to get a #14 arc-welder's glass—those with lower numbers don't filter out enough light. You can find arc-welder's glass at welding supply stores, listed in the Yellow Pages under "Welding Equipment and Supplies." The glass comes in rectangular pieces measuring about 2 by 4 inches and imparts a greenish hue to the Sun. Unfortunately, the glass isn't of high enough optical quality to be used over the front end of a telescope or as a photographic filter. It is best for naked-eye viewing.

If you're interested in photographing the partial phases of an eclipse or viewing the eclipse through a telescope, you'll need to step up to either a Mylar filter or a metal-on-glass filter. Mylar filters consist of one or two sheets of Mylar plastic coated with a thin layer of aluminum. They let more blue light than red light pass through, so they make the Sun appear a cool blue-white color. They are readily available from several companies that supply astronomical accessories, including Celestron International, Thousand Oaks Optical, and Roger W. Tuthill, Inc. Look for these dealers' advertisements in the pages of *Astronomy* magazine or *Sky and Telescope*. If you want a filter for your telescope, the dealers can supply one of the correct size. When you get the filter, check it for defects by holding it up to a bright light. If any light leaks through, the filter probably has a pinhole that would let through too much of the Sun's radiation.

Never use as a solar filter aluminized Mylar sold as "space blankets," window insulation, or candy wrappers. This stuff is far too flimsy and lets through too much light.

The most expensive way to filter sunlight is with a metal-on-glass filter. These filters are specifically designed to fit over the front end of a telescope. Manufacturers deposit a nickel-chromium alloy known as Inconel onto optical-quality glass in a vacuum chamber. (For greater durability, stainless steel is sometimes added to the Inconel.) These filters transmit slightly more red light than blue light, yielding a yellow-orange Sun that many observers find aesthetically more pleasing than the colors other filters produce. The premier manufacturer of metal-on-glass filters is Thousand Oaks Optical. They make over-the-aperture filters—filters that fit over the front end of a telescope or pair of binoculars — for all brands.

WHAT TO AVOID

Knowing which filters are safe for viewing the Sun is only half the knowledge you need. You also must know which filter materials are unsafe. One of the most dangerous devices you'll find is the so-called "Sun filter" included in some small-aperture, imported telescopes. These work on the same principle as over-the-aperture filters do, except that they screw on the bottom of the eyepiece and absorb the light there instead of intercepting the Sun's light before it enters the telescope. The problem is that the telescope's lens or mirror still collects all the Sun's radiation and focuses it near the eyepiece. The "Sun filter" absorbs this intense heat, which can crack the filter and let through the full power of the Sun's amplified light. Permanent blindness then occurs in a fraction of a second. If you own or buy a telescope that comes with an eyepiece Sun filter, throw it away immediately. Never use it to observe the Sun.

Perhaps the most publicized alternative filter is to layer two or more strips of black-and-white film that has been fully exposed and then developed. The problem with this method is that it only works with some black-and-white films. The technique requires a silver-based emulsion—it's the silver left in the negatives that blocks the sunlight. Unfortunately, many newer black-and-white films are not silver-based. Even when fully exposed and developed, their negatives transmit far too much infrared radiation. Unless you know exactly what you're doing, this method is too dangerous. (In addition to the right kind of film, you must use two layers and shouldn't gaze at the Sun for more than about thirty seconds at a time.) Because safe solar filters are inexpensive and easy to find, we recommend avoiding black-and-white film altogether.

Even worse, however, is color film. Color slide film, if left unexposed and then developed, is essentially opaque to visible radiation. But this produces a false sense of security because the film is nearly transparent to infrared radiation. The danger from color slide film is not only that dangerous levels of infrared radiation pass through, but also that the Sun itself appears so dim that you'll be tempted to look for longer periods.

Other filters to avoid include sunglasses (either standard or polarized), smoked glass, crossed polarizing filters, black plastic garbage bags, and X-ray film. Photographic neutral density filters are dangerous too—even those rated ND5. All these materials limit the amount of visible radiation that passes through, but do little or nothing to stop the hazardous infrared.

SOLAR PROJECTION

Experiencing a partial eclipse indirectly by projecting an image of the Sun to view is not only safe but also offers a pleasing image. There are several ways you can do this. The first is to create your own pinhole camera. Start with two pieces of stiff white cardboard and cut a square hole in the center of one of them roughly two inches on a side. A matte knife cuts best, but a pair of scissors also works. Then tape a piece of aluminum foil over the hole. To make the pinhole part of the camera, simply take a straight pin and poke a small hole in the center of the aluminum foil. Rotate the pin slightly to smooth out any rough edges, but keep the hole as small as possible. To observe the eclipse, let the Sun's light pass through the pinhole and onto the second piece of cardboard, which serves as a screen. (It's easiest to use if you stand with your back to the Sun.) Remember—do not look at the Sun directly through the pinhole, but instead at the image of the Sun projected on the screen.

The pinhole camera will form an inverted image of the Sun on the screen. If you hold the screen closer to the pinhole, the image grows brighter; if you move the screen farther away, the image grows larger. By adjusting the separation of the two cardboard pieces, you can get the most appealing combination of image size and brightness.

A pinhole camera is one of the best ways to show an eclipse to a small group of people. You can also make a pinhole camera for one person out of a cardboard box. This makes a great project for children, who will each get a personal view of the eclipse. The pinhole camera also has a built-in safety feature—it keeps a child looking in the direction opposite the Sun.

First, get your hands on an old cardboard box. Because the image of the Sun gets bigger in direct proportion to the length of the box, try to get one 18 inches long or longer. If you don't have a spare box lying around, you can usually find one at your local grocery store. An old shoe box will do in a pinch, but the Sun's image will end up quite small.

Cut a square hole measuring about two inches on a side in the center of one end of the box. Tape a piece of aluminum foil over the hole. Next tape a white card or a piece of white paper on the inside of the opposite end of the box to serve as the projection screen. Then cut a large hole in one of the long sides of the box. The hole should be near the end with the foil-covered hole and should extend about half the length of the box. Leave about an inch of cardboard at both the top and bottom. This hole functions as a window for viewing the image of the Sun.

Next poke a small hole in the aluminum foil with a straight pin or sewing needle, as you did for the large version of the camera. The final step is to tape shut any flaps on the box. You want to keep the inside of the box as dark as you can, so that the Sun's image shows brightly. The only light inside the box should be sunlight coming through the pinhole and a slight amount of

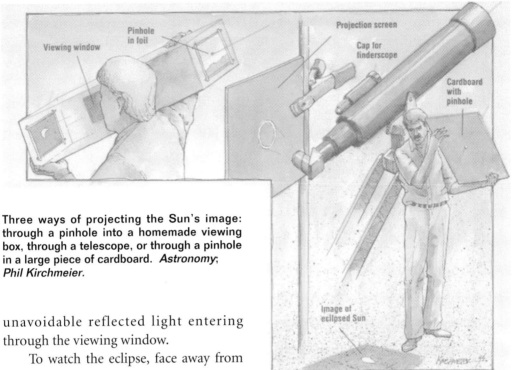

Three ways of projecting the Sun's image: through a pinhole into a homemade viewing box, through a telescope, or through a pinhole in a large piece of cardboard. *Astronomy*; *Phil Kirchmeier.*

unavoidable reflected light entering through the viewing window.

To watch the eclipse, face away from the Sun and hold the box on your shoulder. Point the foil end toward the Sun with the viewing window next to your face and move the box around until you've made its shadow as small as possible. When you look at the screen in the box, you'll see a miniature, inverted image of the eclipsed Sun. Of course you shouldn't wait until eclipse day to test the projection box—try it out beforehand on any sunny day.

The ultimate size of the projected image will be about $^1/_{110}$, or a little less than one percent, of the distance between the pinhole and the screen. That's why using a big box is important—you need a box 11 inches long just to make the Sun's image 0.1 inch tall, and a box 27 inches long to get an image a quarter of an inch tall.

PROJECTING WITH A TELESCOPE

You can also use a telescope or a pair of binoculars to project an image onto a shaded screen. This is the oldest technique for viewing magnified images of the Sun, employed by Galileo and his contemporaries. Although it is a little tricky and requires a bit more care than using an over-the-aperture filter, it provides a large, detailed image of the Sun that many people can view at the same time.

Set up your scope as if you were going to observe the Sun with a filter. Begin by centering the Sun in the field of view by minimizing the shadow cast by the telescope's tube, as we discussed in Chapter 5. As with a pinhole camera, you'll need a white screen on which to project the image. A large piece of cardboard or a taut sheet will do the job nicely. You can even set up an easel with paper on it and sketch the Sun during the partial phases. The projected image will be detailed enough for you to see large sunspots easily.

Use caution when projecting the Sun's image. Before you start, cap the finderscope so no one looks through it. If you use binoculars, project through only one of the optical pairs and keep the other capped. Keep in mind that the projected solar beam can get quite hot. It's best to stay clear of it, especially if you have long hair or a beard. Finally, never leave the telescope unattended—someone could come along and unwittingly try to look through the scope.

Although the safety of the people around you is most important, the health of your telescope is also at risk when you project the Sun's image. When unfiltered sunlight passes through it, heat can build up and damage the optics. That's why you shouldn't use a Schmidt-Cassegrain or Maksutov telescope to project the Sun: The heat that builds up inside the tube can damage the secondary mirror. Whatever telescope you use, stop down its aperture to one to two inches. You can do this simply by cutting a 1- to 2-inch circle out of a piece of cardboard and taping the entire piece over the front of the scope so that the hole lies away from the optical axis. This will keep too much heat from building up inside the telescope. Also, periodically allow the optics to cool down by putting the lens cap back on or throwing a light-colored towel or sheet over the front end of the telescope.

We recommend using a low-power eyepiece in your projection setup. That way the whole Sun gets projected, not just part of it. If you don't have a clock drive on your scope, a low-power eyepiece also means you won't have to re-center the image as often. And don't use your most expensive eyepiece— the heat that builds up in the system can start to melt the cement that holds the various lenses in place.

Few events in astronomy can compare with seeing a solar eclipse, but few are fraught with as much danger. Be careful. Protect your eyes from harmful levels of solar radiation. With a good filter or solar projection, you can witness this breathtaking event in all its glory, and save your eyesight for the next one while you're at it.

A ring of fire sets into the Pacific Ocean during the annular eclipse of January 1992. *Jorge Soto.*

8

ECLIPSES OF THE NEXT DECADE

Eclipse chasers will be able to keep themselves very busy for the rest of the 1990s. Every continent except Antarctica plays host to at least one total or annular eclipse during the remainder of the decade, and most of these eclipses offer exciting destinations, long eclipses, or good weather prospects. Let's take a look at the nine central eclipses between 1994 and 1999 to see what each will deliver.

ANNULAR ECLIPSE OF MAY 10, 1994

The first is an annular eclipse that cuts a wide swath across the heart of North America on May 10, 1994. The Moon's shadow first touches Earth at sunrise well out in the Pacific Ocean, some 450 miles southeast of Hawaii. It then races toward the east-northeast, making landfall along the west coast of Baja California at 15h54m universal time. (Universal time, or UT, is an unambiguous system of reckoning time used by astronomers around the world. Based on a 24-hour clock, UT corresponds to the standard time on the meridian that passes through Greenwich, England. To convert from UT to North American time, subtract 5 hours for EST, 6 hours for CST, 7 hours for MST, and 8 hours for PST. During daylight time, subtract one hour less than these values.)

The shadow's path traverses the central part of the Baja Peninsula, where the annular phase lasts 5 minutes 28 seconds and the Sun appears 40° above the eastern horizon. The shadow then passes over the Gulf of California and strikes mainland Mexico near the town of Bahía Kino. It crosses northwestern Mexico before reaching the southwestern United States. The southeastern tip of Arizona, southern New Mexico, and western and northern Texas will all see an annular eclipse shortly after 10 a.m. MDT. El Paso, Texas, lies near the center line and experiences 5 minutes 40 seconds of annularity, beginning at 16h07m UT (10:07 a.m. MDT).

From the Texas panhandle, the Moon's shadow heads northeast through central Oklahoma, southeastern Kansas, and central Missouri, passing over Oklahoma City and St. Louis and just south of Kansas City. From St. Louis, which lies near the southern limit of the path of annularity, the eclipse starts at 16h52m UT (11:52 a.m. CDT) and lasts 3 minutes 44 seconds with the Sun 65° above the southeastern horizon. Compare that with a point along the center line in Missouri, where the annular phase lasts over 6 minutes.

Crossing the Mississippi River into Illinois, the path neatly bisects the Land of Lincoln before heading into northern Indiana. Both Chicago and Indianapolis just miss seeing the annular eclipse, witnessing instead a partial eclipse that covers 94 percent of the Sun.

Continuing east and slightly north, the path then heads into northern Ohio and southern Michigan. Maximum eclipse occurs just west of Toledo, Ohio, where annularity lasts 6 minutes 13 seconds and the Sun lies 66° above the southern horizon at 17h13m UT (1:13 p.m. EDT). At that point the Moon covers 94 percent of the Sun's diameter and 89 percent of the Sun's area. All of Lake Erie and Lake Ontario, as well as their U.S. and Canadian shorelines, lie within the path. Inhabitants of Detroit, Cleveland, Buffalo, and Toronto will all see an annular eclipse.

The track then heads through upstate New York and northern New England, covering almost all of Vermont and New Hampshire and the southern half of Maine. It then leaves the United States and heads back into Canada, grazing the southern coast of New Brunswick before encompassing most of Nova Scotia. From Halifax, the annular eclipse starts at 17h56m UT (2:56 p.m. AST) and lasts 5 minutes 53 seconds with the Sun 55° high in the southwest.

But the eclipse isn't over yet. After traversing the North Atlantic, the path of annularity once again touches land in the Azores, where the eclipse lasts 5 minutes 10 seconds and the Sun lies 27° high. The path finally reaches the Atlantic coast of Morocco as sunset approaches. Residents of Casablanca will witness an annular eclipse lasting 4 minutes 33 seconds with the Sun poised just 3° above the ocean.

Eclipse chasers have a number of good spots from which to witness annularity. The best weather prospects along the entire track appear to be in eastern Baja California and western mainland Mexico, where you can expect clear skies 80 to 90 percent of the time in May. (Avoid the west coast of Baja, where fog and low clouds are common on May mornings.)

For those who prefer viewing the eclipse from the United States, the southwest offers the best weather prospects. Arizona, New Mexico, and southwest Texas promise a 65 to 75 percent chance of clear skies at eclipse time. As the eclipse path heads into the Great Plains, prospects for clear skies drop significantly. Since moist air from the Gulf of Mexico dominates the Midwest at this time of year, this region offers no better than a 50-50 chance of good weather.

As the eclipse approaches its maximum duration in Ohio, prospects dim further—to about 40 percent. Heading east into New England and the Canadian Maritimes, the odds of clear weather drop to just 1 in 3. Despite the relatively poor prospects in these regions, however, all is not lost. If a cold front moves through the region shortly before eclipse day, the weather could be gorgeous, with clear, deep-blue skies.

Throughout the United States and Canada the observer has the advantage of good weather forecasts. Begin following the forecast five days before the eclipse. As eclipse day approaches, keep an eye on the forecast and on weather satellite images to pinpoint the spots most likely to have clear skies. And keep mobile—often a few hours' drive on eclipse morning will get you to a clear area.

Prospects for clear weather at the end of the eclipse in Africa don't look all that great. Casablanca is a fairly cloudy city where the weather is clear only about 30 percent of the time during late afternoons in May. If you opt for viewing the eclipse from Africa, you're probably best off heading slightly inland. The slopes of the Atlas Mountains offer about a 50 percent chance for a sunny sunset.

TOTAL ECLIPSE OF NOVEMBER 3, 1994

Totality next touches Earth in South America, where the Sun will disappear for between three and four minutes on the morning of November 3, 1994. The Moon's shadow first touches down in the eastern Pacific, nearly 1,000 miles west of Peru. Flying quickly over the ocean waters, the shadow comes ashore along the southern coast of Peru near that country's border with Chile and Bolivia. Right at the northern limit of the path lies Nazca, where the Sun will briefly go out over the huge, ancient geometric patterns of birds and animals etched in the desert.

After hugging the coast for over 100 miles, the track finally heads inland near Mollendo, Peru, which lies 60 miles southwest of Arequipa, Peru's second-largest city. Here totality lasts just under three minutes with the Sun appearing 29° above the eastern horizon, beginning at 12h15m UT.

The path of totality then heads east-southeast across the Atacama Desert and Andes Mountains of southern Peru and northern Chile before reaching the altiplano of southern Bolivia. The altiplano is a wide, high plateau some 12,000 feet above sea level. On the center line of the eclipse, just south of Lago de Poopó, the Sun will be in eclipse for a bit over three minutes, starting at 12h22m UT.

The wealth of sunspots visible during the partial phases of the July 1991 eclipse likely won't be repeated in 1995, when sunspot activity should be at low levels. *Mark Coco.*

From there the eclipse path heads over the eastern Andes and moves onto lower ground in Paraguay. From Asunción, the capital, which lies near the southern limit of the eclipse track, the Moon will cover the Sun for just cover the Sun for just 50 seconds. On the center line north of the capital, the Sun will be gone for a more respectable 3 minutes 37 seconds.

As the Moon's shadow reaches Paraguay's border with Brazil and Argentina, it passes over one of the most scenic spots on the globe: Iguaçú Falls. These spectacular falls along the Paraná River are 100 feet taller and four times wider than Niagara Falls on the United States-Canadian border. Here totality lasts 3 minutes 31 seconds, commencing at 12h46m UT with the Sun 52° high.

From the falls, the path heads across southern Brazil, leaving the continent near the city of Criciúma. This point offers the longest eclipse on land, 4 minutes 1 second of totality, beginning at 12h59m UT with the Sun 59° above the east-northeastern horizon. Although Brazil marks the last landfall of the eclipse path, maximum eclipse occurs in the South Atlantic at latitude 35° south and longitude 34° west. From there the Sun will be hidden for 4 minutes 23 seconds beginning at 13h40m UT.

Eclipse chasers concerned only with the prospects for good weather will flock to the western part of South America, away from the coast. Even though the Atacama Desert is one of the driest regions in the world, clouds are common right along the coast, particularly in the morning when the Moon's shadow comes calling. Better to head a bit inland, where you quickly gain altitude and rise above the marine layer.

Chances for clear skies are just 30 percent along the coast, but jump to 70 to 80 percent when you climb to an altitude of just 3,000 feet. (Note that roads in this desert region are few and far between.) The chances for clear weather are even better in the Bolivian altiplano: between an 80 and 90 percent chance of clear weather on eclipse day. Be warned, however, that at the altitude of the altiplano, air contains only 60 percent of the oxygen it does at sea level. If you plan on observing from this region of Bolivia, arrive several days ahead of time so you can get acclimatized.

If the length of totality is more important to you than the weather prospects, then the more accessible lowlands of Paraguay, Argentina, and Brazil may be for you. Here you'll get up to about four minutes of totality, but you'll have only about a 50-50 chance of clear weather. If you opt for viewing the eclipse from eastern South America, keep mobile so you can travel to where clear skies are most likely.

ANNULAR ECLIPSE OF APRIL 29, 1995

One of the least appealing eclipses this decade is the annular eclipse of April 29, 1995. Although this is the longest eclipse remaining in the twentieth century, with the annular phase lasting 6 minutes 37 seconds at maximum, it occurs over mostly inaccessible areas of northern South America.

The path of annularity first strikes Earth far out in the South Pacific, then rushes northeast toward the northern coast of Peru. It comes ashore near the Peru-Ecuador border in the early afternoon, plunging the city of Piura, Peru, into near darkness. There the annular phase starts at 17h24m UT and lasts 6 minutes 21 seconds with the Sun 70° above the northern horizon. Maximum eclipse occurs just 100 miles northeast of Piura, high up in the Andes Mountains, where the Moon will hide 95 percent of the Sun's diameter.

After crossing mostly uninhabited regions of northern Peru and southern Colombia, the track heads nearly due east across the equally uninhabited and inaccessible Amazon River basin of northern Brazil. The only cities of note in the path of annularity are Belém and Fortaleza, both located along the northeast coast of Brazil. In Belém the annular phase starts at 19h03m UT and lasts 5 minutes 22 seconds with the Sun lying 30° above the western horizon. Annularity lasts 4 minutes in Fortaleza, beginning at 19h15m UT, and the Sun's altitude has dropped to 17°.

Remote location and poor weather prospects for this eclipse make it one of the least anticipated

If you head to the April 1995 annular eclipse in northern South America, expect to see the Moon cover just a bit more of the Sun's disk than it did in January 1992. *Jorge Soto.*

eclipses of the decade. The Amazon rain forest, as you might guess, sees frequent clouds and rain, and April is still near the peak of its rainy season. Your best bet for clear weather is in western South America, along the slopes of the towering Andes.

TOTAL ECLIPSE OF OCTOBER 24, 1995

The next total eclipse cuts a narrow swath across Asia on October 24, 1995. Although it is a fairly short eclipse, with the Moon covering the Sun for just over two minutes at maximum, weather prospects are good in many of the areas.

The eclipse begins as the Sun rises in Iran, some 50 miles due south of Tehran, and lasts only 19 seconds just as the Sun pokes above the eastern horizon. The eclipse track then heads southeast across Iran, Afghanistan, and Pakistan before heading into India. Calcutta is the one major city that lies in the path of totality in India. From this metropolis the Sun will disappear for 1 minute 22 seconds, beginning at 3h19m UT, when our star lies 40° above the southeastern horizon.

From India, the path skims the southern coast of Bangladesh, then cuts through Myanmar, Thailand (passing slightly north of Bangkok), Cambodia, and Vietnam. In Cambodia, the eclipse track passes over the twelfth-century ruins at Angkor Wat, which should make a fascinating vista for viewing the eclipse. Few major cities lie in the path of totality simply because the path is so narrow, reaching a maximum diameter of just 50 miles.

After leaving the coast of Southeast Asia, the eclipse track heads across the South China Sea, touching the northern tip of the island of Borneo at 4h45m UT. Maximum eclipse takes place slightly north of Borneo, when the Sun disappears for 2 minutes 10 seconds. No other major islands lie in the path of totality, which continues far out into the Pacific Ocean.

This is one eclipse where weather prospects may have to take a back seat to political concerns. With the eclipse path passing through such early 1990s political hot spots as Iran, Afghanistan, and Myanmar, you had best keep apprised of the latest goings-on as eclipse time approaches. (See Chapter 4 for tips on how you can come by this information.)

As for the weather prospects, the best chances for clear weather should be near the India-Pakistan border, where the skies over the Great Indian Desert

Exposing once every ten minutes for an hour recorded both the partial phases and totality of the February 1979 solar eclipse. *William P. Sterne, Jr.*

are clear 90 percent of the time on October mornings. The Calcutta area also looks promising, with clear skies occurring 60 to 70 percent of the time. As you head farther east, the prospects deteriorate. Expect only about a 30 to 40 percent chance of clear weather as you head into Southeast Asia and the islands of the South China Sea.

TOTAL ECLIPSE OF MARCH 9, 1997

After two partial eclipses in 1996, the next total eclipse arrives on March 9, 1997. This may be the least exciting total eclipse of the 1990s. Not only is it a fairly short eclipse, it lies in a cold, remote part of the globe. Only a few hardy eclipse chasers will follow the Moon's shadow this time.

The umbra touches down at sunrise near the western border of Mongolia and Russia. From there it heads due east, hugging the border between those countries and skirting the southern shore of Lake Baikal. The track then grazes the northeastern corner of China before veering to the northeast. After crossing central Siberia, the Moon's shadow plunges into the Arctic Ocean before leaving Earth not far from the North Pole. Maximum eclipse occurs at 1h24m UT in Siberia, when the Sun will disappear for 2 minutes 50 seconds and lie 23° above the southeastern horizon.

Weather prospects for this eclipse are actually pretty good if you pay attention only to the possibility of cloud cover. Along the early part of the eclipse track in northern Mongolia and southern Russia, chances are about 50-50 for clear skies on eclipse day. As you head closer to the site of maximum eclipse, the probability of clear weather improves to 70 to 80 percent. Unfortunately, early March is still the dead of winter in Siberia: Average daily highs this far north reach only -5° Fahrenheit and overnight lows average a bone-chilling -40° F.

TOTAL ECLIPSE OF FEBRUARY 26, 1998

Just under 12 months after the cold Siberian eclipse comes a warm and probably clear Caribbean event. The February 26, 1998, eclipse promises nicer and more exotic locales for eclipse observers, ranging from islands in the Pacific and Caribbean to mainland South America.

After traversing a good portion of the eastern Pacific, the Moon's shadow first touches land in the Galápagos Islands, some 750 miles off the coast of Ecuador. The eclipse path doesn't make a direct hit on the Galápagos—it clips the northern edge of the island group. The center line remains north of the islands, but the northern part of Isla Isabela and the small islands of Isla Marchena and Isla Pinta will see the total phase. Totality occurs in late morning, beginning at 16h57m UT, and lasts 2 minutes 42 seconds with the Sun 68° high in the east-southeast. On the center line north of the islands, totality will be a healthy 4 minutes 4 seconds long.

The eclipse track then heads to the northeast, reaching land again where Central America joins South America. (Maximum eclipse occurs in the Pacific between the Galapagos and mainland South America, where the Moon blocks the Sun for 4 minutes 9 seconds and the Sun is 76° high.) The path then crosses the southern tip of Panama, northern Colombia, and northwestern Venezuela before moving into the Caribbean Sea. The biggest city in the path of totality is Maracaibo, Venezuela, where the Sun disappears from the sky for 2 minutes 56 seconds, beginning at 18h04m UT and is then 65° high in the southwest.

The Moon's shadow makes landfall several times in the Caribbean, first of all on the islands of Aruba and Curacao in the Netherlands Antilles. You can expect over 3 minutes of totality on these islands, beginning at about 18h10m UT. Then after a long stretch over water, the path of totality hits land again in the Leeward Islands. Antigua, Guadeloupe, and Montserrat are the favored islands, where totality will last about 3 minutes 20 seconds on the center line and the Sun will be 50° high. The track then heads off across the Atlantic, leaving Earth at sunset just west of the Canary Islands.

Overall weather prospects for this eclipse appear reasonably good. The Galapagos tend to be cloudy more often than not, so you should expect only a 40 percent of clear skies at eclipse time. By the time the eclipse track hits South America, however, the odds of good weather increase to about 50 percent. The best spot on the mainland for eclipse viewing should be north of Maracaibo, where February is the height of its dry season and early afternoon is the brightest part of the day.

Even though the islands of the Caribbean are mainly dry, convective clouds occasionally build up late in the day, which is the time of the eclipse. The clouds should, however, dissipate as the air cools between first and second contact. You can anticipate sunny skies and good eclipse weather 50 to 60 percent of the time on February afternoons.

ANNULAR ECLIPSE OF AUGUST 22, 1998

When the Moon next passes across the center of the Sun, its apparent diameter is not large enough to cover the entire disk, so we see an annular eclipse. The ring of sunlight can be seen on August 22, 1998, from many of the islands lying between Asia and Australia.

The Moon's shadow begins its journey across Earth's surface at sunrise in the Indian Ocean. The track heads east from there, first hitting land in Sumatra and then in Malaysia. From Sumatra, the annular phase starts at 0h19m UT and lasts 2 minutes 48 seconds, but the Sun is only 15° above the eastern horizon. Continuing eastward, the Moon's shadow then crosses northern Borneo, where the ring of fire lasts 10 seconds longer than in Sumatra and the Sun has climbed to an altitude of 33°.

The large size of the Moon's disk relative to the Sun's, and the shadow's path across the tropics combined to make the July 1991 eclipse particularly long. Eclipse chasers of the next decade will have to cram their observations into a shorter schedule. *Jim Curry.*

After Borneo, most of the rest of the path of annularity lies over open water. Although the shadow manages to hit a few small islands, it avoids most of the larger ones. The path just misses the northern edge of the island of New Guinea, where maximum eclipse occurs some 100 miles offshore at 2h06m UT. Ninety-seven percent of the Sun's disk is hidden for 3 minutes 14 seconds and the Sun lies 75° up in the northern sky. The last major landfall comes on the island of New Britain, just east of New Guinea, where the annular phase begins at 2h24m UT and continues for 2 minutes 59 seconds with the Sun 72° high.

Unfortunately, weather in this part of the globe is often cloudy, so it could be hard to get a good view of annularity. Clouds are more common along the western edge of the track. From Sumatra and Borneo you can expect only a 30 to 40 percent chance of clear weather on eclipse morning, but on the island of New Britain the chances improve to roughly 50 percent.

ANNULAR ECLIPSE OF FEBRUARY 16, 1999

Just six months after the August 1998 annular eclipse, another one takes place not too far away. Although the eclipse track starts south of Africa, it traverses mostly open water in the Indian Ocean until reaching Australia. That's where eclipse chasers will head to view this one. And with the Moon covering a bit over 99 percent of the Sun during this eclipse, there should be a fine showing of Baily's beads.

The path of annularity crosses the sparsely settled interior of the island continent, making its way through Western Australia, the Northern Territory, and northern Queensland. When the eclipse track first comes ashore some 200 miles north of Perth, annularity lasts a brief 47 seconds, beginning at 7h30m UT, with the Sun 45° high in the west. By the time the Sun is in eclipse over Queensland at 8h05m UT, the annular phase lasts 1 minute 5 seconds and the Sun has sunk to an altitude of only 16°.

Weather prospects look reasonably good for this eclipse. February is summertime in Australia, and temperatures tend to be hot and skies clear. You can expect about a 70 percent chance of clear weather over interior sections to watch for Baily's beads rimming the limb of the Moon.

TOTAL ECLIPSE OF AUGUST 11, 1999

The last central eclipse of the century also promises one of the better shows. The total eclipse of August 11, 1999, passes through much of Europe and into Asia. Sophisticated weather forecasts and easy access to observing sites help balance the poorer weather prospects for this eclipse. And it should occur around the time of the next peak in solar activity, so the corona should appear nice and round.

Although Europe and Asia are the prime observing sites, the Moon's shadow first contacts Earth in the western Atlantic, some 200 miles south of Nova Scotia. The shadow avoids land until it crosses the North Atlantic,

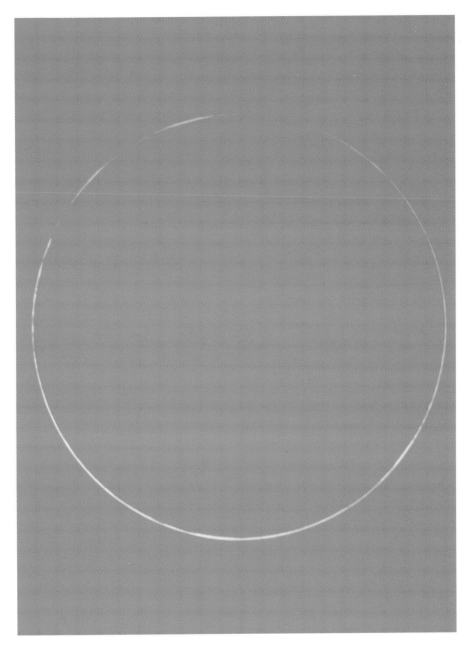

The May 1984 annular eclipse came about as close to being total as possible, leaving a paper-thin ring of sunlight highlighted by numerous Baily's beads. *Jim Curry.*

however, making landfall in southwestern England. At Falmouth, which lies near the center line, the Sun will lie 46° above the horizon when it disappears for 2 minutes 3 seconds, starting at 10h11m UT.

After plunging the Cornwall coast into darkness, the shadow heads across the English Channel and into northern France. Paris lies just outside the path of totality, so Parisians will have to travel north several miles or be content with a 99 percent partial eclipse. The eclipse track then passes over southern Belgium and Luxembourg before heading into southern Germany, where both Stuttgart and Munich will see darkness at midday. At Munich, totality begins at 10h37m UT and continues for 2 minutes 7 seconds, as the Sun appears 56° above the south-southeastern horizon.

After cutting through central Austria, including Salzburg, the eclipse track heads through Hungary, northern Serbia, and Romania. Bucharest lies right on the center line and will experience 2 minutes 22 seconds of totality, starting at 11h06m UT.

Totality leaves Europe near the Romania-Bulgaria border as the Moon's shadow heads across the Black Sea and into Asia. The umbra then crosses central Turkey, northern Syria and Iraq, central Iran, southern Pakistan, and central India before leaving Earth at sunset in the Bay of Bengal.

If you plan on observing this eclipse from Europe, you'll want to keep up with the latest forecasts and stay mobile, so that you can pick out a spot likely to have clear skies and then be able to get there. Western England has about a 50 percent chance of good weather on eclipse day, which rates a little higher than most of western and central Europe. But the best weather prospects in Europe are along the Black Sea coasts of Romania and Bulgaria, where you can expect clear skies about 70 percent of the time.

As you head into Asia you run into even better weather, particularly over the deserts of Iraq and Iran—where you'll have a 90 percent chance of nice weather on eclipse day. Of course, the potential hazards of visiting this region may outweigh your desire for guaranteed good weather, particularly when Europe offers good weather forecasts and easy access to possible observing sites. Once you get into Pakistan and India, the weather prospects dim once more, as the wet summer monsoons are in full swing. Expect only a 30 percent chance of nice weather during the late afternoon hours.

That summarizes all the total and annular eclipses through the end of the

The twentieth century concludes with a spectacular total eclipse visible over much of Europe. The corona should look much as it did during the February 1979 eclipse, when the Sun was also very active. *Alan Dyer.*

decade. The year 2000—technically the last year of the decade and century—features four eclipses, but all are partials. Not until June 2001 will we witness another central eclipse, and by then we should have a new edition of this book to keep you up to date!

AFTERWORD

We have surveyed the geometry, history, significance, and recording of solar eclipses. These events have made a profound impact on mankind throughout history. From the time of the ancients until today, people have been fascinated by solar eclipses perhaps more than any other natural event. Only a few centuries ago, people were beset by feelings of foreboding in the event of an eclipse. An eclipse was a harbinger of bad things: floods, war, and famine. Such events often meant it was time for whoever was in charge of the local government to get out of town, fast.

Today the opposite is true. People flock to the site of a total solar eclipse. Consider the eclipse of July 11, 1991, which more people observed than any other in history. Somewhere in the vicinity of 30 million people saw that particular eclipse, and of those, at least five percent apparently made a planned, concerted effort to get into the path of totality.

With few exceptions, eclipses are probably the only dramatic, powerful, uncontrollable phenomenon in nature that humans are inexorably drawn to. With few exceptions, most similar natural events are dangerous. You don't see this many people spending money and flocking to see hurricanes, volcanos, or earthquakes! If the past five years are any indication, even greater numbers of amateur astronomers and other interested people will join in chasing the shadow. But at least until the end of this century, most residents of North America, Australia, and Africa will have to travel some distance to see totality. This will separate the serious eclipse chaser from the once-in-a-lifetime participant.

There will probably be even more companies offering expeditions to the eclipse experience. Following the 1991 event, there has been an explosion of firms billing themselves as "experienced," "professional," or "expert" in organizing and operating eclipse trips. Here we caution, "let the buyer beware." You may think the worst thing that can happen to you in the care of an inexperienced tour company is to miss the eclipse, but far worse outcomes could await tour members who connect with a negligent or careless tour operator. Protect yourself by asking many detailed questions. If the operator seems upset or annoyed by your inquisitiveness, seek another.

Are there any untapped locations or methods of observing solar eclipses the common person has yet to employ? Two possibilities immediately come to mind, if you use a bit of imagination. One is the observation of annularity or totality from Earth orbit, either from the space shuttle or an orbiting space station. Unfortunately, anyone seeking to observe an eclipse from one of these platforms had better be prepared for a short experience. Assuming the vehicle travels at an orbital velocity of 17,500 miles per hour and intercepts the largest possible lunar shadow at the best possible angle, the eclipse would last just 34.2 seconds—hardly worth all the effort and expense.

The second scenario would be to watch a solar eclipse from the best vantage point in our solar system: the Moon. Think of it—no moisture, no air, no possibility of being clouded-out. And with Earth subtending an apparent diameter nearly four times as big as the Moon in our sky, a solar eclipse viewed from the Moon would last as much as 90 minutes or more.

Remember the geometry here. A total solar eclipse as seen from the Moon is the same event as a total lunar eclipse as seen from Earth. It would be the ultimate "fix" for the most hard-core eclipse addict. On top of the perfect conditions for viewing the event, the huge terrestrial disk, in combination with the airless environment of the Moon, would allow observers to see the solar corona stretch dozens of millions of miles into space—farther than anyone has ever viewed it from here on our planet. True, essentially all solar prominences would be blotted out by the huge Terran disk, but this would be a small price to pay to witness totality from this magnificent perspective.

Eclipse chasers of the next few centuries will very likely look back on our primitive efforts with amusement. We look back on the gallant eclipse pursuers of the eighteenth and nineteenth century in the same way, tempered by our wonder at their endurance of barbaric conditions en route. Following that centuries-old standard, we can only look on our own efforts as one of the most exciting chapters of our lives, driven by awe and wonder. Go for it!

APPENDIX A

CENTRAL ECLIPSES: 2001-2020

LOCAL CIRCUMSTANCES AT GREATEST ECLIPSE

JUNE 21, 2001

Type	Magnitude	Longitude	Latitude	Sun's altitude	Sun's azimuth
Total	1.050	2°44.6' E	11°15.8' S	55°	354°

The twenty-first century gets off to a great start with a long total eclipse for the Southern Hemisphere. Totality begins at sunrise in the South Atlantic, some 200 miles east of the mouth of the Río de la Plata. From there the path of totality heads northeast and then, after passing 0° longitude, heads east-southeast across southern Africa. There it traverses central Angola, central Zambia, northern Zimbabwe, and central Mozambique before crossing the Mozambique Channel and southern Madagascar. Totality lasts 4 minutes 57 seconds at maximum in the South Atlantic, but is still greater than 3 minutes throughout Africa.

DECEMBER 14, 2001

Type	Magnitude	Longitude	Latitude	Sun's altitude	Sun's azimuth
Annular	0.968	130°41.4' W	0°37.3' N	66°	189°

The path of annularity for this eclipse passes mainly over the great expanse of the central Pacific, making little landfall. However, after avoiding most of the islands in the Pacific (missing the Hawaiian Islands by a couple hundred miles), the path does cross Central America. Observers in northern Costa Rica and southern Nicaragua will witness an annular eclipse late in the afternoon that lasts up to 3 minutes. People in much of North America can see a partial eclipse covering up to 60 percent of the Sun.

JUNE 10, 2002

Type	Magnitude	Longitude	Latitude	Sun's altitude	Sun's azimuth
Annular	0.996	178°36.7' W	34°32.7' N	78°	170°

It's hard to imagine an eclipse path that covers 7,000 miles and yet never hits land, but this one comes very close. Starting in the Celebes Sea just north of the Indonesian island of Sulawesi, the path of totality heads northwest across the Pacific Ocean, avoiding all but a few tiny islands. Then just as the Sun begins to set, the Moon's shadow briefly touches the west coast of Mexico. Viewers near Puerto Vallarta will see an annular eclipse lasting about 1 minute. Even though no one will experience much annularity, many will see a good partial eclipse. Observers in California will see better than 60 percent of the Sun covered at maximum, while others in western North America will see anywhere from 20 to 80 percent coverage.

DECEMBER 4, 2002

Type	Magnitude	Longitude	Latitude	Sun's altitude	Sun's azimuth
Total	1.024	59°33.6' E	39°28.2' S	72°	15°

A short eclipse late in 2002 brings totality to two continents. Shortly after sunrise on December 4, the Moon's shadow swoops across southern

Africa. The shadow first strikes land along the coast of central Angola—the exact same area that saw totality just 18 months earlier. (On average an individual location on Earth's globe sees totality once every 300 years, so the residents of coastal Angola really beat the odds at the beginning of this century!) From Angola the track of totality hugs the border between Zimbabwe and Botswana, then nips northern South Africa and southern Mozambique before heading into the Indian Ocean. By late afternoon the Moon's shadow reaches southern Australia, not too far north of Adelaide. The length of totality near both sunrise and sunset is a little under 1 minute.

MAY 31, 2003

Type	Magnitude	Longitude	Latitude	Sun's altitude	Sun's azimuth
Annular	0.938	24°21.4' W	66°37.7' N	3°	35°

A far northern eclipse provides over 3 minutes of annularity to observers in Iceland. This unusual eclipse features a path of annularity nearly 2,800 miles wide—by far the widest path for any central eclipse this century. Unfortunately, at such high latitudes, the huge width of the path doesn't translate into having lots of people living in the path of annularity.

NOVEMBER 23, 2003

Type	Magnitude	Longitude	Latitude	Sun's altitude	Sun's azimuth
Total	1.038	88°18.3' E	72°39.3' S	15°	111°

Six months after an annular eclipse in the far north, Earth experiences a total eclipse in the far south. The path of totality touches land only in Antarctica, starting at sunrise due south of the Cape of Good Hope then continuing eastward in a swath some 300 miles wide. Maximum eclipse lasts for nearly 2 minutes but occurs in the frozen wasteland several hundred miles from the Antarctic coast.

APRIL 8, 2005

Type	Magnitude	Longitude	Latitude	Sun's altitude	Sun's azimuth
Annular-total	1.007	118°57.7' W	10°34.6' S	70°	331°

A rare annular/total eclipse takes place in April 2005 when the Moon's umbral shadow is just long enough to reach Earth's surface at maximum eclipse but not long enough to reach the surface at the beginning or end of the eclipse. Thus people near the end of the track witness an annular eclipse while those near the center see a total eclipse. Most of the track for this eclipse traverses the open waters of the Pacific, but landfall does occur in Latin America late in the afternoon. Residents and eclipse chasers in parts of Panama, northern Colombia, and northern Venezuela will see an annular eclipse lasting up to 30 seconds.

OCTOBER 3, 2005

Type	Magnitude	Longitude	Latitude	Sun's altitude	Sun's azimuth
Annular	0.958	28°45.0' E	12°52.4' N	71°	210°

Because about 75 percent of Earth's surface is covered with water, most eclipses spend more time over water than land. But the October 2005 eclipse is a notable exception. The Moon's shadow first touches Earth in the North Atlantic, but then quickly makes its way to the Iberian Peninsula. After bringing near darkness to northern Portugal and central Spain (Madrid experiences the gloom for just over 4 minutes), the shadow heads across the Mediterranean and into Africa. There it traverses northeast Algeria, central Tunisia, Libya, the northeastern tip of Chad, the Sudan, southwestern Ethiopia, northern Kenya, and southern Somalia. Maximum eclipse occurs in central Sudan, where the annular phase lasts 4 minutes 31 seconds.

MARCH 29, 2006

Type	Magnitude	Longitude	Latitude	Sun's altitude	Sun's azimuth
Total	1.052	16°45.9' E	23°08.6' N	67°	149°

Africa also plays host to the next solar eclipse, a long total eclipse in March 2006, but other areas get in on the action too. Early morning risers slightly north of Recife, Brazil, will see the totally eclipsed Sun rise out of the Atlantic. After a long trip across the Atlantic, the Moon's shadow next strikes land in southern Ghana, Africa, and then heads northeast across Togo, Benin, northwest Nigeria, Niger, northwestern Chad, Libya (crossing some of the same spots to see annularity six months earlier), and the northwestern tip of Egypt. But it's not through yet. After crossing the Mediterranean, the path bisects Turkey before hitting Kazakhstan and ending in Russia. Eclipse chasers along the Chad-Libya border will see the longest eclipse of 4 minutes 7 seconds.

SEPTEMBER 22, 2006

Type	Magnitude	Longitude	Latitude	Sun's altitude	Sun's azimuth
Annular	0.935	9°03.4' W	20°39.9' S	66°	31°

A long annular eclipse follows six months later, but only people on boats or in planes will see the maximum eclipse of slightly over 7 minutes. The only land-based sites in the path of annularity are in northern South America. Residents of and visitors to the northern halves of Guyana, Suriname, and French Guiana will see a ring of sunlight in the early morning hours of September 22 that lasts up to 5 minutes.

FEBRUARY 7, 2008

Type	Magnitude	Longitude	Latitude	Sun's altitude	Sun's azimuth
Annular	0.965	150°26.9' W	67°34.6' S	16°	269°

After a lull of over a year, the next central eclipse comes in February 2008. Unfortunately, like the November 2003 eclipse, only people and penguins in Antarctica will see the Sun in eclipse. The path of annularity cuts through Ellsworth Land before leaving the frozen continent and heading into the Amundsen Sea. The maximum eclipse of 2 minutes 12 seconds takes place some 1,500 miles off the coast.

AUGUST 1, 2008

Type	Magnitude	Longitude	Latitude	Sun's altitude	Sun's azimuth
Total	1.039	72°16.3' E	65°38.2' N	34°	236°

Slightly more hospitable conditions, though not significantly less remote, await eclipse chasers in August 2008. This is the next total eclipse visible from North America, though just the northernmost parts of Canada will see totality. The eclipse begins at sunrise on Victoria Island, then heads northeast across Prince of Wales Island, Somerset Island, Devon Island, and Ellesmere Island. After hugging the northern Greenland coast, the path skirts the north pole and then plunges south into Russia, western Mongolia, and northern China. At its maximum in northern Russia, totality lasts 2 minutes 27 seconds.

JANUARY 26, 2009

Type	Magnitude	Longitude	Latitude	Sun's altitude	Sun's azimuth
Annular	0.928	70°16.4' E	34°04.7' S	73°	336°

A long annular eclipse begins a twelve-month period that includes the two longest annular eclipses and the longest total eclipse during the years 2001 to 2020. The January 26, 2009 eclipse starts at sunrise in the Atlantic southwest of the Cape of Good Hope. From there the Moon's shadow heads northeast across the Indian Ocean, where it reaches a maximum duration of 7 minutes 54 seconds. The Moon's shadow continues on, however, making landfall in the Greater Sunda Islands—southern Sumatra, western Java, and Borneo—before leaving Earth at sunset in the Celebes Sea. Observers along the center line in southern Sumatra will see 6 minutes of annularity.

JULY 22, 2009

Type	Magnitude	Longitude	Latitude	Sun's altitude	Sun's azimuth
Total	1.080	144°08.4' E	24°12.0' N	86°	201°

The longest total eclipse in this twenty-year period comes in July 2009 and can be seen from eastern Asia. The eclipse begins in western India, just north of Bombay. From there it heads nearly due east, encompassing the southern tip of Nepal, the northern tip of Bangladesh, and most of Bhutan before crossing southern China. Eclipse observers in Shanghai will witness a total eclipse lasting over 5 minutes. Once off the mainland and into the East China Sea, the Moon's shadow races across a few of the small Ryukyu Islands before going on to reach maximum eclipse not far into the western Pacific. Anyone who takes a cruise to this spot will see a spectacular eclipse lasting 6 minutes 39 seconds.

JANUARY 15, 2010

Type	Magnitude	Longitude	Latitude	Sun's altitude	Sun's azimuth
Annular	0.919	69°20.2' E	1°37.3' N	66°	166°

Like the long annular eclipse of January 2009, this annular eclipse peaks in the middle of the Indian Ocean. But unlike that previous eclipse, this one lasts significantly longer (a whopping 11 minutes 8 seconds at maximum — the longest in the twenty-first century) and offers more land access. The eclipse begins as the Sun rises in the Central African Republic, then quickly proceeds across northern Zaire, Uganda, Kenya, and southern Somalia. Here annularity lasts around 8 minutes. After crossing the Indian Ocean, the shadow darkens the southeastern coast of India and northern Sri Lanka. The Moon's shadow next crosses the Bay of Bengal and then the country of Myanmar before heading into China. Observers along the center line in China can see up to 8 minutes of annularity. The Moon's shadow finally leaves Earth in the Yellow Sea, just short of the Korean Peninsula.

JULY 11, 2010

Type	Magnitude	Longitude	Latitude	Sun's altitude	Sun's azimuth
Total	1.058	121°51.0' W	19°46.5' S	47°	13°

Another lengthy eclipse takes place in July 2010, but this one is mostly over water. At maximum in the South Pacific, the Sun will disappear completely for 5 minutes 20 seconds. This point lies well away from land, however, so your best bet may be to take a cruise from one of the islands in French Polynesia. If you want to stay on land, consider viewing the eclipse at sunset from southern Chile or southern Argentina. From there the eclipse lasts a little less than 3 minutes.

MAY 20, 2012

Type	Magnitude	Longitude	Latitude	Sun's altitude	Sun's azimuth
Annular	0.944	176°19.1' E	49°04.6' N	61°	172°

This is the first central eclipse since the annular one on May 10, 1994, that will be visible from the United States. The eclipse begins on the other side of the Pacific, however, in southern China. Shortly after sunrise, residents of Hong Kong will see a ring of sunlight lasting 3 minutes 30 seconds. The path of annularity then traverses the East China Sea and envelops southern Japan at midmorning, leaving Tokyo in partial darkness for just over 5 minutes. Finally, after making its way across the North Pacific, the Moon's shadow comes ashore on the Oregon-California border. The path of annularity then sweeps through southern Oregon, northern California, Nevada, southern Utah, northern Arizona, the southwestern tip of Colorado, New Mexico, and western Texas, where the eclipse ends at sunset. The maximum length of annularity in the United States is nearly 5 minutes along the Oregon-California coast.

NOVEMBER 13, 2012

Type	Magnitude	Longitude	Latitude	Sun's altitude	Sun's azimuth
Total	1.050	161°17.9' W	39°57.6' S	68°	11°

Totality returns to Australia in November 2012, but to the remote northern part of the island-continent. Just after sunrise on the 14th, the Sun is completely hidden from view by our satellite for residents and eclipse chasers in the northern part of the Northern Territory just east of Darwin and in the central sections of the Cape York Peninsula in Queensland. Here totality lasts a respectable 2 minutes. From Australia the track heads across the South Pacific, where the eclipse reaches a maximum duration of just over 4 minutes, but makes no other significant landfall.

MAY 10, 2013

Type	Magnitude	Longitude	Latitude	Sun's altitude	Sun's azimuth
Annular	0.954	175°30.5' E	2°12.3' N	74°	349°

Six months later the Moon's shadow returns to Australia, though this time the eclipse is annular and not total. Beginning at sunrise in Western Australia, the track heads to the northeast across the Northern Territory and then into Queensland, encompassing some of the same parts of the Cape York Peninsula graced by totality the previous November. The maximum duration for annularity in Australia is over 4 minutes. From Australia, the track grazes the southeastern edge of Papua New Guinea and then passes over the Solomon Islands, the last major landfall for this eclipse.

NOVEMBER 3, 2013

Type	Magnitude	Longitude	Latitude	Sun's altitude	Sun's azimuth
Annular/ total	1.016	11°39.6' W	3°29.4' N	71°	193°

The other side of the world sees the next eclipse, an annular-total affair in November 2013. Starting as an annular eclipse in the North Atlantic, it quickly becomes total as the Moon's shadow heads southeast. The track passes less than 100 miles south of Liberia, where the eclipse reaches a maximum duration of 1 minute 40 seconds. The path then swings more eastward and intercepts the African coast in central Gabon. From there the track passes through the Congo, northern Zaire, northern Uganda, northern Kenya, and southern Ethiopia, where the eclipse turns back to an annular one before ending at sunset.

APRIL 29, 2014

Type	Magnitude	Longitude	Latitude	Sun's altitude	Sun's azimuth
Annular	0.986	131°09.5' E	70°41.8' S	1°	319°

Perhaps the most unusual of all the eclipses in this twenty-year span is the annular eclipse of April 2014. It is essentially an instantaneous eclipse, meaning annularity lasts only a split second and is visible from only one spot on the globe. And because that spot lies in the interior of Antarctica, it's quite likely that no human will witness this eclipse.

MARCH 20, 2015

Type	Magnitude	Longitude	Latitude	Sun's altitude	Sun's azimuth
Total	1.045	6°34.1' W	64°25.0' N	19°	135°

A somewhat more appealing eclipse arrives 11 months later, but it's not what you would call a must-see event. The total eclipse of March 2015 traverses the North Atlantic and Arctic Oceans, not exactly garden spots during late winter. Denmark's Faeroe Islands and the Norwegian island of Svalbard offer the only land access. Maximum eclipse occurs slightly north of the Faeroe Islands and lasts for 2 minutes 47 seconds.

MARCH 9, 2016

Type	Magnitude	Longitude	Latitude	Sun's altitude	Sun's azimuth
Total	1.045	148°50.2' E	10°06.5' N	75°	164°

After two frigid eclipses comes a welcome relief: A long, warm weather total eclipse. The Moon's shadow first touches Earth in the Indian Ocean, where the Sun disappears in eclipse just as it pushes above the horizon. The track then heads east, first crossing Sumatra, then Bangka and Belitung, southern Borneo, central Sulawesi, and Halmahera. On these Indonesian islands, totality lasts about 4 minutes. To reach the full duration of 4 minutes 10 seconds, however, you'll have to take a cruise to the central Pacific.

SEPTEMBER 1, 2016

Type	Magnitude	Longitude	Latitude	Sun's altitude	Sun's azimuth
Annular	0.974	37°48.1' E	10°41.4' S	71°	16°

Central and southern Africa play host to an annular solar eclipse in September 2016. The track begins in the South Atlantic, but the Moon's shadow quickly pushes to the coast of Gabon. It heads east-southeast from there, traversing the Congo, central Zaire, eastern and southern Tanzania, and northern Mozambique. After crossing the Mozambique Channel, the Moon's shadow cuts through northern Madagascar and Reunion before lifting back into space. Maximum for this eclipse occurs in southern Tanzania, where the annular phase lasts 3 minutes 6 seconds.

FEBRUARY 26, 2017

Type	Magnitude	Longitude	Latitude	Sun's altitude	Sun's azimuth
Annular	0.992	31°08.4' W	34°41.6' S	63°	340°

The first solar eclipse of 2017 is a short annular eclipse visible from parts of South America and Africa. The Moon's shadow first encounters Earth more than 1,000 miles off the southwestern coast of South America. From there it moves quickly northeastward and slices through southern Chile and southern Argentina. The shadow continues on its course and by late afternoon reaches southern Africa, where it traverses central Angola and northern Zambia before leaving Earth's surface. The maximum duration for this eclipse is about 1 minute in both South America and Africa.

AUGUST 21, 2017

Type	Magnitude	Longitude	Latitude	Sun's altitude	Sun's azimuth
Total	1.031	87°37.7' W	36°58.0' N	64°	199°

This is the eclipse observers in the United States have been waiting for. It's the first total solar eclipse to hit part of the continental United States since February 1979. Although it offers only 2 minutes 40 seconds of totality at maximum in western Kentucky, everyone on the center line will see the Sun disappear for at least about 2 minutes. The eclipse track cuts across the United States from northwest to southeast, passing through good chunks of Oregon, Idaho, Wyoming, Nebraska, Missouri, Illinois, Kentucky, Tennessee, and South Carolina before heading out into the Atlantic.

July 2, 2019

Type	Magnitude	Longitude	Latitude	Sun's altitude	Sun's azimuth
Total	1.046	108°57.0' W	17°23.5' S	50°	359°

Although the total eclipse of July 2019 occurs mostly over water, it does offer eclipse chasers a few spots for good observing. The vast majority of the eclipse track lies in the South Pacific, where the length of totality reaches a maximum of 4 minutes 33 seconds. But the eclipse track ends up in South America, passing through north-central Chile and Argentina. In Chile the track encompasses La Serena, where the eclipse lasts just over 2 minutes. In Argentina the Sun sets while in total eclipse from just west of the capital, Buenos Aires.

December 26, 2019

Type	Magnitude	Longitude	Latitude	Sun's altitude	Sun's azimuth
Annular	0.970	102°18.4' E	0°59.7' N	66°	184°

Southern Asia is the site for the annular eclipse of December 2019. Starting at sunrise on the Arabian Peninsula, the eclipse track cuts across Saudi Arabia, Bahrain, the United Arab Emirates, and Oman, bringing about 3 minutes of annularity to observers on the center line. The path next heads across the Arabian Sea to southern India and northern Sri Lanka, where eclipse chasers can expect about 3 minutes 15 seconds of annularity. Then the track crosses the Bay of Bengal and slices through Sumatra, Malaysia, and Borneo before heading out to the open seas of the western Pacific. Maximum eclipse occurs in eastern Sumatra, where annularity lasts 3 minutes 39 seconds.

JUNE 21, 2020

Type	Magnitude	Longitude	Latitude	Sun's altitude	Sun's azimuth
Annular	0.994	79°43.2' E	30°31.0' N	83°	177°

Just six months after the December 2019 eclipse, another annular eclipse follows a similar but slightly more northerly course. This eclipse begins a bit farther west, however, in northern Zaire. The early morning part of the path traverses the eastern tip of the Central African Republic, southern Sudan, and Ethiopia, where eclipse observers can witness a bit over 1 minute of annularity. After crossing the Red Sea, the southern part of the Arabian Peninsula, and the Gulf of Oman, the Moon's shadow passes over Pakistan, northern India, Tibet, southern China, and Taiwan before leaving Earth in the western Pacific.

DECEMBER 14, 2020

Type	Magnitude	Longitude	Latitude	Sun's altitude	Sun's azimuth
Total	1.025	67°54.4' W	40°21.1' S	73°	9°

The final eclipse of the first 20 years of the twenty-first century is a rather short total eclipse visible from part of South America. The eclipse track passes over Chile and Argentina, from near Valdivia on the Chilean coast to the Golfo San Martías on the Argentine coast. The maximum for this eclipse occurs in south-central Argentina, where eclipse chasers can see 2 minutes 10 seconds of totality.

PARTIAL ECLIPSES: 1995-2020

LOCAL CIRCUMSTANCES AT GREATEST ECLIPSE

Date	Magnitude	Longitude	Latitude	Area covered
Apr. 17, 1996	0.879	103°57.8' W	71°20.8' S	New Zealand, Antarctica
Oct. 12, 1996	0.757	32°09.1' E	71°42.5' N	northeast Canada, Greenland, Europe, northern Africa
Sep. 2, 1997	0.898	114°12.9' E	71°46.0' S	Australia, New Zealand, Antarctica
Feb. 5, 2000	0.579	134°10.5' E	70°13.1' S	Antarctica
Jul. 1, 2000	0.476	109°27.5' W	66°55.9' S	southern South America
Jul. 31, 2000	0.603	59°50.8' W	69°31.7' N	Russia, Greenland, northwestern North America
Dec. 25, 2000	0.723	74°06.3' W	66°20.2' N	North America, Central America
Apr. 19, 2004	0.735	44°20.6' E	61°35.6' S	southern Africa, Antarctica
Oct. 14, 2004	0.928	153°36.6' W	61°15.0' N	northeastern Asia, western Alaska
Mar. 19, 2007	0.874	55°24.7' E	61°02.9' N	Asia, Alaska
Sep. 11, 2007	0.749	90°17.4' W	61°00.3' S	South America, Antarctica
Jan. 4, 2011	0.858	20°48.8' E	64°39.1' N	Europe, northern Africa, western Asia
Jun. 1, 2011	0.602	46°49.4' E	67°47.0' N	northern Asia, northern North America, Greenland
Jul. 1, 2011	0.096	28°38.9' E	65°09.5' S	southern Ocean
Nov. 25, 2011	0.904	82°24.0' W	68°34.1' S	southern Africa, Antarctica, Tasmania, New Zealand
Oct. 23, 2014	0.811	97°04.6' W	71°10.1' N	eastern Asia, North America
Sep. 13, 2015	0.787	2°18.4' W	72°06.4' S	southern Africa, Antarctica
Feb. 15, 2018	0.598	0°44.5' E	71°01.6' S	Antarctica, southern South America
Jul. 13, 2018	0.336	127°28.2' E	67°55.4' S	southern Australia
Aug. 11, 2018	0.736	174°32.9' E	70°23.3' N	Greenland, northern Europe, northern Asia
Jan. 6, 2019	0.715	153°37.0' E	67°26.1' N	northeastern Asia

APPENDIX B

ECLIPSE MAPS

MAY 1994

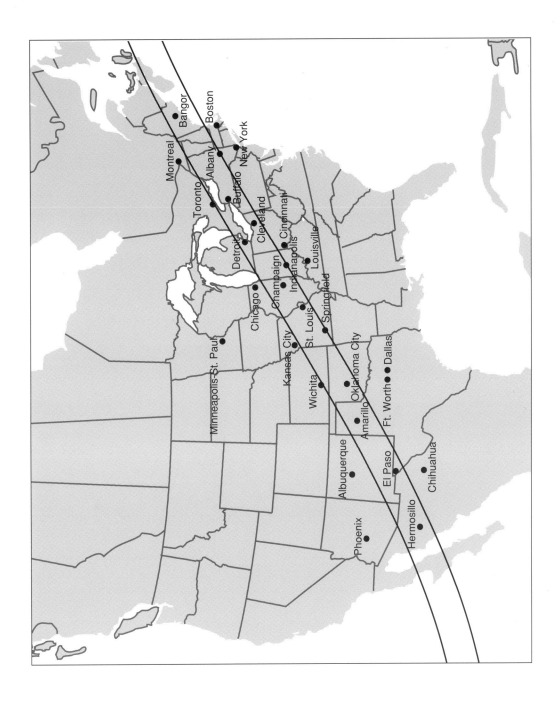

NOVEMBER 1994

APRIL 1995

OCTOBER 1995

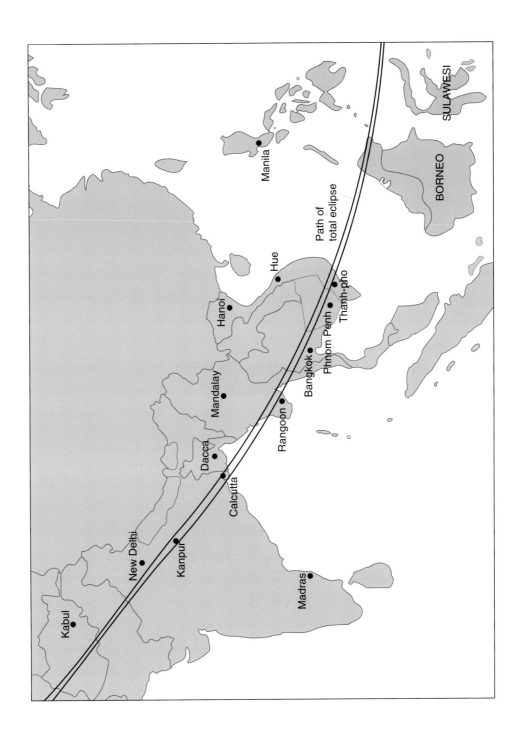

MARCH 1997

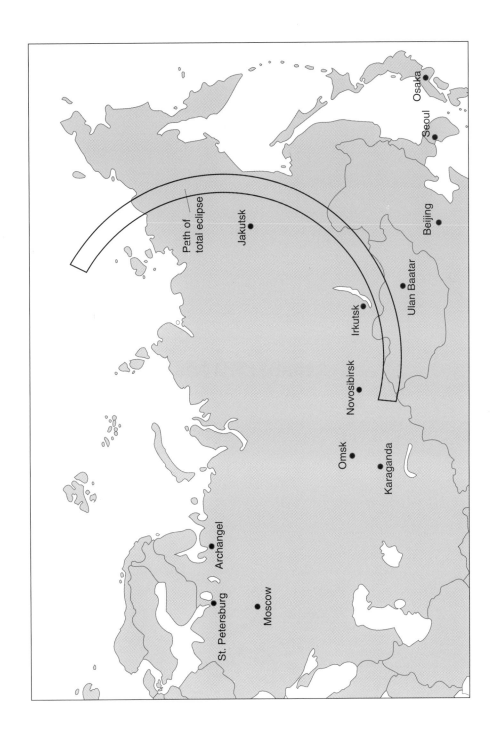

FEBRUARY 1998

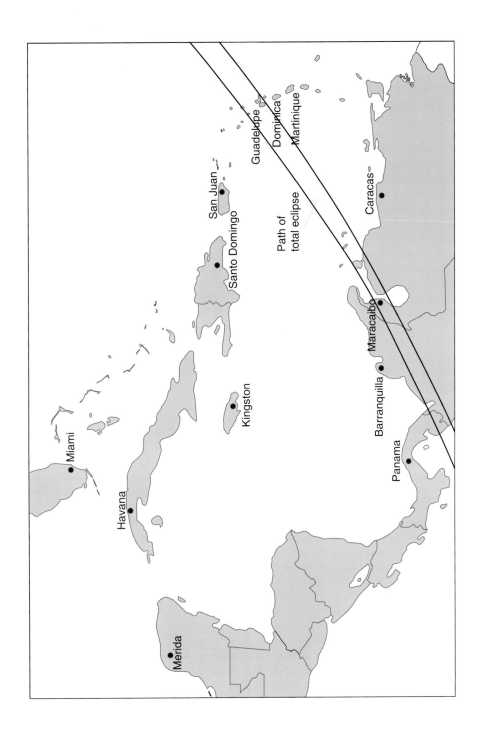

AUGUST 1998

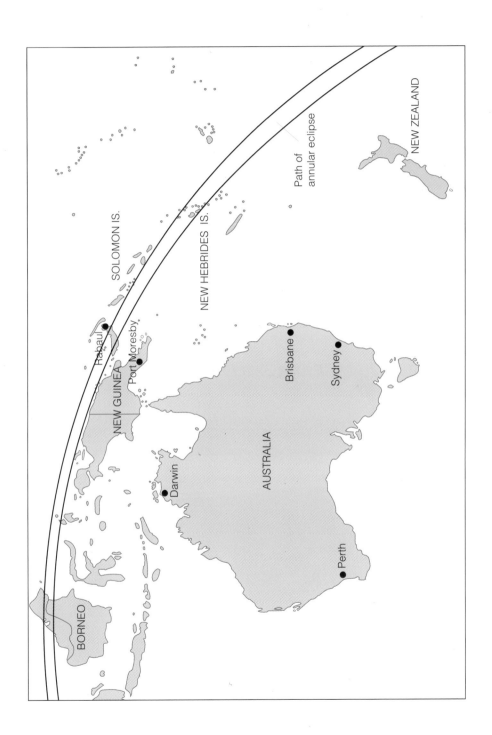

FEBRUARY 1999

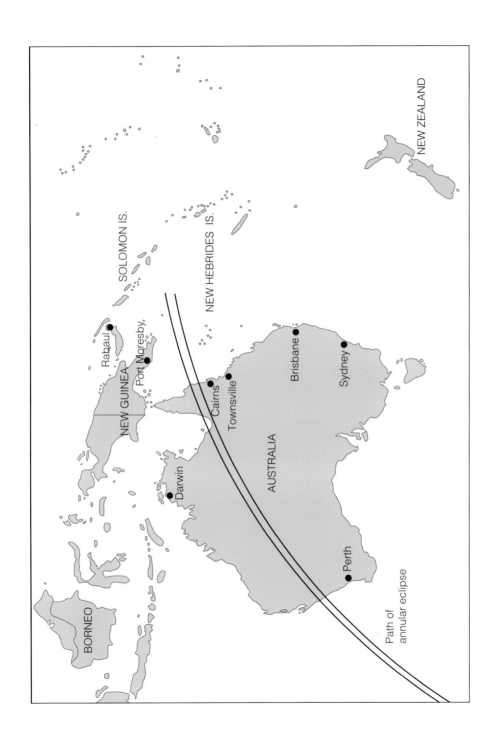

AUGUST 1999

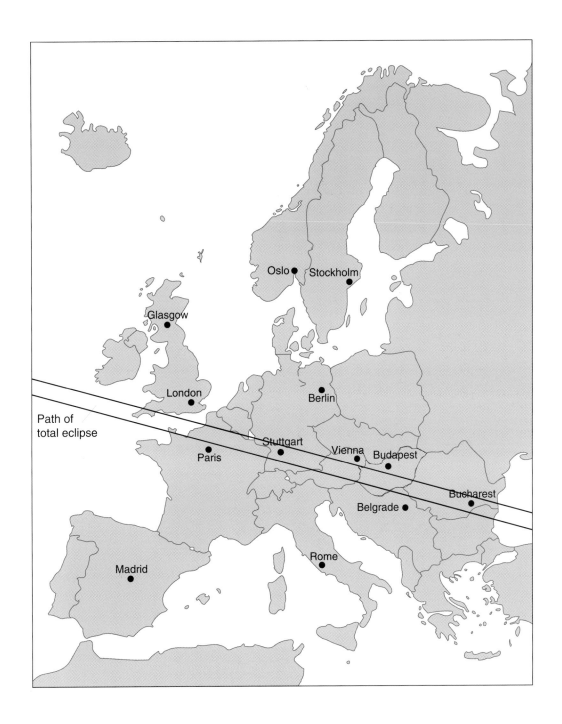

STAR MAPS

The maps on the following pages depict the sky as it will appear during totality for each of the five total solar eclipses remaining this decade. All are wide-field views showing the planets and stars brighter than 3rd magnitude, about the limit of what can be seen with the naked eye during totality.

NOVEMBER 3, 1994

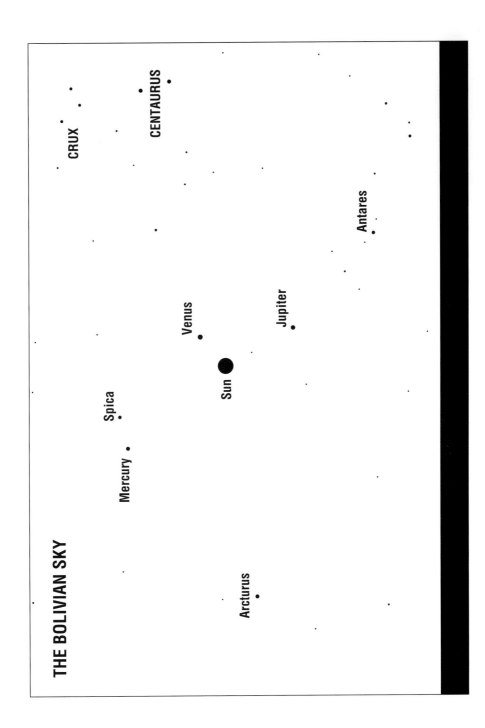

OCTOBER 24, 1995

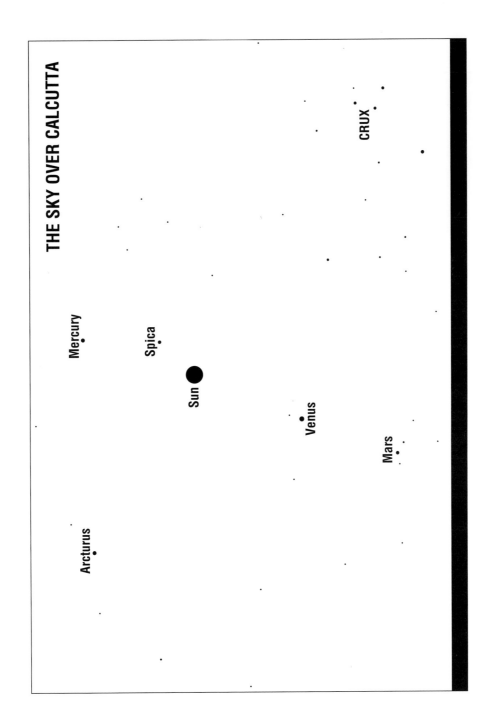

THE SKY OVER CALCUTTA

Arcturus
Mercury
Spica
Sun
Venus
Mars
CRUX

MARCH 9, 1997

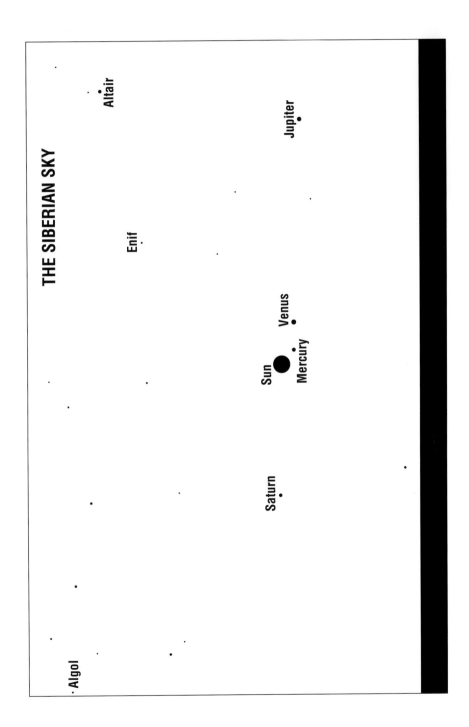

THE SIBERIAN SKY

Algol

Altair

Enif

Jupiter

Sun
Venus
Mercury

Saturn

FEBRUARY 26, 1998

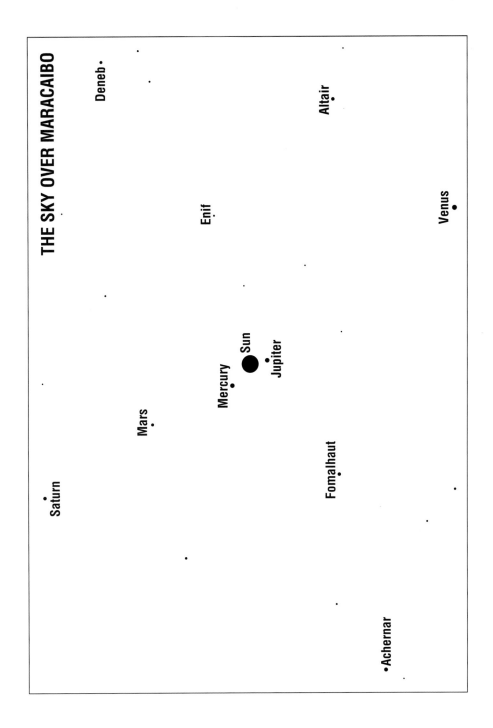

THE SKY OVER MARACAIBO

AUGUST 11, 1999

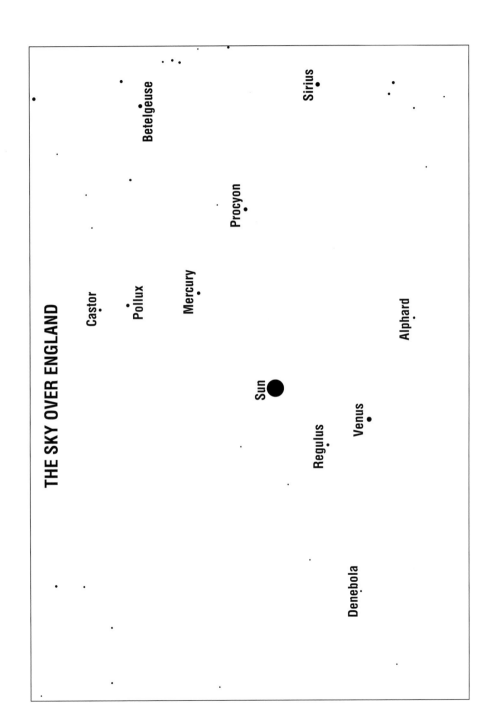

THE SKY OVER ENGLAND

BIBLIOGRAPHY

Allen, David and Carol. *Eclipse.* North Sydney, Australia: Allen and Unwin, 1987.

Astronomical Almanac for the Year 1994. Nautical Almanac Office, U. S. Naval Observatory, Washington, D.C., 1993.

Astrophotography Basics. Kodak Publication No. P-150, Eastman Kodak, Rochester, New York, 1988.

Brewer, Bryan. *Eclipse.* Second edition. Seattle, Washington: Earth View, 1991.

Covington, Michael. *Astrophotography for the Amateur.* Cambridge, England: Cambridge University Press, 1991.

Espenak, Fred. *Fifty Year Canon of Solar Eclipses: 1986–2035.* NASA Reference Publication 1178. 1987 revision.

Espenak, Fred, and Jay Anderson. *Annular Solar Eclipse of 10 May 1994.* NASA Reference Publication 1301, 1993.

Espenak, Fred, and Jay Anderson. *Total Solar Eclipse of 3 November 1994.* NASA Reference Publication 1318, 1993.

Explanatory Supplement to the Astronomical Almanac. Edited by P. Kenneth Seidelmann. Mill Valley, California: University Science Books, 1992.

Fiala, Alan D., James A. DeYoung, and Marie R. Lukac. *Solar Eclipses, 1991–2000.* Circular No. 170, Nautical Almanac Office, U. S. Naval Observatory, Washington, D.C., 1986.

Littmann, Mark, and Ken Willcox. *Totality, Eclipses of the Sun.* Honolulu: University of Hawaii Press, 1991.

Meeus, Jean. *Astronomical Formulae for Calculators.* Second edition. Richmond, Virginia: Willmann-Bell, 1982.

Meeus, Jean. *Astronomical Tables of the Sun, Moon, and Planets.* Richmond, Virginia: Willmann-Bell, 1983.

Meeus, Jean, Carl Grosjean, and Willy Vanderleen. *Canon of Solar Eclipses.* Oxford, England: Pergamon Press, 1966.

Mucke, Hermann, and Jean Meeus. *Canon of Solar Eclipses, -2003 to +2526.* Second edition. Vienna, Austria: Astronomisches Büro, 1983.

Oppolzer, Theodor von. *Canon of Eclipses.* New York: Dover Publications, 1962.

Pasachoff, Jay M., and Michael A. Covington. *The Cambridge Eclipse Photography Guide.* Cambridge, England: Cambridge University Press, 1993.

Smart, William M. *Textbook on Spherical Astronomy.* Sixth edition. Cambridge, England: Cambridge University Press, 1977.

Stephenson, F. Richard, and M. A. Houlden. *Atlas of Historical Eclipse Maps, East Asia 1500 BC–AD 1900.* Cambridge, England: Cambridge University Press, 1986.

Zirker, Jack B. *Total Eclipses of the Sun.* New York: Van Nostrand Reinhold, 1984.

ENDNOTES

1. E. C. Krupp, interview by author, Griffith Observatory, December 21, 1992.

2. P. M. Muller and F. R. Stephenson, *The Accelerations of the Earth and Moon from Early Astronomical Observations,* G. D. Rosenberg and S. K. Runcorn, (London: John Wiley, 1975), p. 459.

3. Ibid., 477.

4. F. R. Stephenson and Said S. Said, "Precision of Medieval Islamic Eclipse Measurements," M. A. Hoskin, *Journal for the History of Astronomy* (Bucks: Science History Publications Ltd, 1991), p. 197.

5. Said S. Said, F. R. Stephenson, and Wafiz Rada, "Records of Solar Eclipses in Arabic Chronicles," *Bulletin of the School of Oriental and African Studies* (London: University of London, 1989), p. 51.

6. E. C. Krupp, *Beyond the Blue Horizon* (New York: Harper Collins, Inc., 1991), p. 163.

7. Ibid., 162–163.

8. J. E. Kennedy, "Media Coverage of the Solar Eclipse Expedition of 1860 to Northern Canada," *The Musk-Ox,* Shirley M. Milligan (Department of Geological Sciences, University of Saskatchewan, 1986), p. 46.

9. Ibid., 48.

10. Ibid, 46.

11. S. A. Newcomb, personal diary, July 20, 1860, University of Saskatchewan Archives.

12. E. A. Maunder, *The Total Solar Eclipse—1900* (London: British Astronomical Association, Witherby and Co., 1901), p. 36.

13. Edward Maunder, *The Story of Eclipses* (New York: McClure, Phillips, and Co., 1904), p. 100.

14. George F. Chambers, *The Story of Eclipses* (New York: McClure, Philips, and Co., 1904), p. 141.

15. Isabel M. Lewis, *A Handbook of Solar Eclipses* (New York: Duffield and Company, 1924), p. 4.

16. Ibid., 93.

INDEX